NIGHT HAUNTS

NIGHT HAUNTS

Sukhdev Sandhu

Λrtangel

VERSO
London • New York

First co-published by Artangel and Verso 2007

© Sukhdev Sandhu 2006

The moral right of the author has been asserted

Artangel

31 Eyre Street Hill, London EC1 5EW

www.artangel.org.uk

1 3 5 7 9 10 8 6 4 2

Verso

UK: 6 Meard Street, London W1F 0EG

USA: 180 Varick Street, New York, NY 10014-4606

www.versobooks.com

Verso is the imprint of New Left Books

ISBN-13: 978 1 84467 162 5

British Library Cataloguing in Publication Data

A catalogue record for this book is available from the British Library

Library of Congress Cataloging-in-Publication Data

A catalog record for this book is available from the Library of Congress

Typeset in Vaudeville by Eggers + Diaper, Berlin

Printed and bound in Great Britain by
Marston Book Services Limited, Oxfordshire

For – and because of – Jagir Kaur and
Puran Singh Sandhu

CONTENTS

I

An Appetite for Stories:

Introduction

AN APPETITE FOR STORIES:

INTRODUCTION

Whatever happened to the London night?

There was a time, well over a century ago now, when it was considered one of the finest Victorian inventions. Before then, the onset of darkness had spelled an end to the day. It represented its outer limits, its polar extremes. The night was seen as lawless, foreign territory teeming with rogues and banditos who took advantage of what Shakespeare called its 'vast, sin-concealing chaos' to revel in an orgy of depravity and moral pestilence. It snuffed out the civility and social etiquettes of daytime and brought back trace memories of an older London dense with eldritch forestry.

Gas lighting opened up the night. It made the dark city navigable. More than that, it made the darkness itself visible, inspiring etchers and penny-dreadful illustrators to set about delivering chiaroscuric variations on the theme of shadow, gloaming and umbrage to their patrons and editors. London, fat on imperial wealth, was booming and expanding as never before: shops stayed open later, a newly-established police force was able to patrol its streets, under- and overground trains connected the residents of its previously fragmented boroughs.

Brimming with confidence, full of self-love, the capital developed an appetite for stories, both triumphant and harrowing, about the picturesque characters who populated its nocturnal by-ways and crevices. The London night was a homegrown Africa that on-the-make writers scrambled to map and colonize.

Off they trooped: chin-whiskered moral reformers hoping to edify and save the wretched poor huddled in freezing coal sheds in Shadwell; antsy investigative journos-turned-urban mythopoets snouting for florid stories by moving between the masked ball-goers and theatrical aristocracy of the West End and, Hyde to its Jekyll, the East End of lantern-jawed street fighters, idiosyncratic music-hall performers and white lassies hunkered up in Limehouse opium dens run by their slant-eyed Celestial husbands.

And so it went on. As if to refute Antoine de Saint-Exupéry's later invocation of "Night, the beloved. Night, when words fade and things come alive", nocturnal London spawned a growing library of monographs ranging from the circumspect and sociological to the lubricious and hysterical. The city became part docudrama, part carnival of the grotesque. It was a hive of fascination and to it came a steady flow of gawkers, boulevardiers, solitaires, rubberneckers, slummers and sex tourists exercising their newly-found right to roam. Even crime did not stop them: the Ripper murders of the late 1880s led fresh hordes of tourists, many of them all the way from the United States, to hop aboard double-decker buses in order to see at first hand those that Jack London called "the people of the abyss".

The Blitz did for the London night. It produced life-threatening fear rather than flaneurial frissons. A select few foxtrotted in the face of the imminent apocalypse, many more descended into the London Tube to sleep along station platforms, but the streets were blackened for the first time in over a century and the air thick with the acrid smoke of torched humans and devastated factories and buildings. Much of the East End, that hallowed no-go zone of otherness, was bombed to rubble. Hundreds of thousands of Londoners deserted the city in the following decades: they had had enough of darkness.

Stories of young men and women being chased down Wardour Street by buzzbombs, like those of Whitechapel tube sleepers waking up to find morning commuters stepping gingerly over them, can still be heard; but, as with reminiscences of being allowed to help lamplighters on their rounds or of the child-feasting bogeymen who prowled neighbourhoods at night, they are likely to be heard only in London's nursing homes and day centres for the elderly.

It's rare nowadays to hear anyone talk about 'night time in London'. That phrase, and its suggestion of a distinct, cordoned-off territory in which we may immerse ourselves in strange possibilities or make ourselves susceptible to off-kilter enchantments, seems rather old-fashioned. It has been emperilled by New Labour's vision of London – a blinging, pigeon-free, glass-fronted, private-finance-initiative-funded, cappuccino-sipping, Barcelona-mimicking, Euro-piazza festooned, *Vanity Fair*-endorsed, live-forever, things-can-only-get-better fantasia. The city in recent years has witnessed a bevy of real-estate moguls, foreign investors and film directors trading in a slicked-up form of commodity urbanism; equally, the 'London night' has morphed into, and been rebranded as, 'London nightlife'.

Now that most of its factories and workshops have shut down, to be revived occasionally as subjects of melancholy-suffused social history conferences or converted into niche museums for young schoolchildren who would prefer to be eyeballing David Beckham at Madame Tussauds, the capital has embraced its status as a post-industrial hub in which leisure and tourism are sovereign. Fun – its conception, manufacture and promotion – occupies hundreds of thousands of people; it is no longer primarily a much-anticipated evening reward for day-time graft and pen-pushing. Sex, which lewd and bawdy 17th-century nocturnal travellers regarded as the capital's chief attraction, is remotely accessible to anyone with a laptop at all hours of the day.

Night London is endlessly studied and written about – not for any mysteries it may hold, but because it is now seen as an economic unit. It's a potential cash cow for struggling boroughs eager to hold on to some of the money that their residents travel to Camden and Hoxton and Leicester Square to spend. The language of their reports is a flat-footed anti-poetry studded with allusions to strategic guidance, mechanism development and positive visions. Acronyms clog the pages – TfL, EMZs, the latter standing for Entertainment Management Zones, a new term that describes areas in which large numbers of young people like to hang out in the evening.

Cost-benefit analyses are drawn up to wring maximum revenue streams from this new gold dream of a 24/7-capital. Geographic Information System technology is deployed with a view to stoking local regeneration. Street lights are installed and a lot more CCTV cameras. In the end, 'night life' turns out to mean a clutch of surly-bouncer-fronted clubs pumping out monotonous bpm, and from which puke-breathed likely lads emerge at 2 am to pick fights with girls who won't go home with them and Somali cabbies who, reluctant to have their back seats daubed in beer and kebabs, won't take them home.

The power of night has been waning for decades. Astronomers at the Royal Observatory at Greenwich stopped conducting research there as early as the 1950s because the light pollution was too great; few children growing up in the city today will ever have had the privilege of looking up and picking out a lucky star. In fact, the only stars most of us will get to see these days are those awarded by Michelin critics and proudly displayed by stratospherically expensive West End restaurants.

The BBC, which used to stop broadcasting around midnight (its final chorus of *God Save The Queen* carrying the message that royalty was now instructing listeners to go to bed), is now a multi-channel cathode-ray companion throughout the night. The growing number of freelancers, part-timers and casualeers means that the witching hours are populated by more and more people ringing up call centres in Bangalore to check on their insurance premium payments, popping over to the local internet café to do a spot of photocopying, pounding the treadmill at the gym.

Even postmen, who used to go out on their first round in darkness, little torches appended to their heads so that they could read envelope addresses, have had their working hours restructured so that they only deliver mail during daylight hours. No longer will they tread the streets and climb the tower blocks in near-Stygian gloom to bring glad tidings and break-up letters to the breakfast tables of teary-eyed Londoners. No longer will they venture out under thundering skies so menacing that they feel sure they would never see in another morning. No longer do they have the option of nipping out of the sorting office to grab a fag and watch the dawn come up.

But how true are these laments?

It certainly *feels* to me as if the London night has been decommissioned and that its fissile, threatening energies are now spent. However, that's only an instinct: it really needs to be tested out. And that, during the travels narrated in this book, is exactly what I'm going to be doing. I'm keen to reactivate the largely-dormant Victorian and early-twentieth-century genre of the midnight traipse across the metropolis. I will be journeying – from dusk until dawn; by road, air and water; from its concrete centre to its pastoral fringes; from its subterranean cellars to high up in the sky – with a view to exploring the extent to which nocturnal London can counterbalance an increasingly legible and emulsified diurnal London. I want to mainline its covert and shadowy energies, to locate the ghosts of old London that have been unmoored during the makeovers and transformations of recent history, to have adventures with some of the characters who patrol or run amok through its vast acreage.

While airspace and waterways are some of the terrains through which I'll be moving, my main interest is in the streets of the capital. My chief piece of equipment is a pair of sturdy boots. Walking, according to Rebecca Solnit, is a means of uniting heart and head, both emotional and analytical intelligence: "Walking allows us to be in our bodies and in the world without being made busy by them. It leaves us free to think without being wholly lost in our thoughts." Those thoughts, perhaps unduly shaped by other chroniclers of the night, are what I'm worried about; I'm keen to move beyond a Dore-Dickens-Brassai style survey of representations of nocturnal London and, instead, to provide a series of snapshots of its present-day reality. I want to hear what the people who inhabit London at night make of this accelerated, deregulated city, and not just in terms of hardship and brutalising work, but in terms of joy, beauty, ghosts, religion. A lexicon they know, but are rarely encouraged to deploy.

For all that, I do intend to have a companion on these travels: H. V. Morton. Or, strictly speaking, a slightly battered copy of his once-bestselling metrologue *The Nights of London* (1926). He was a beat – not Beat – journo, a hack's hack from Moseley who had the outsider's enthusiasm for the capital and was capable of bashing out articles about it in next to no time – at his peak, he published five books on the topic in barely eighteen months. He saw London as a gloriously crowded theatre stage, a treasure trove of exotic material that he was ready to yomp for many miles to reach. In his later years, he would use the proceeds from his most famous book, a motoring pastoral called *In Search of England* (1927) to retire to South Africa where he would while away the sweltering days playing with model soldiers and writing letters to friends about the awfulness of the blacks in his adopted country and those that had started arriving in Britain after 1948.

But *The Nights of London*, a collection of some of his reports for the Daily Express, shows Morton at his best, pounding the back alleys and wearing out shoe leather as he hoofs it across London to music halls, dodgy pubs, Chinese New Year celebrations; visits zoos and hospitals and the river police at 2am; catches the last omnibus. He marries journalistic precision to dreamy speculation. Not for him the self-obsessed maunderings of psychogeographic writing; he is happy and eager to talk to working Londoners who furnish him with grounded insights that it would be impossible for him to glean on his own. He is droll, occasionally patronising, but always in thrall to his shifting streetscapes. His book, though by no means the last to be written on the topic of London at night, and its Eastward ho! voice is probably unsustainable these days, is the one to which I am most wedded.

During the course of this book I will be revisiting some of the sites that Morton portrayed so vividly as well as moving across newer nocturnal topographies. I will be flying above the city with military helicopters equipped with night-vision cameras that can pick out shirt labels from a height of over 2000 feet; wading through vast rivers of congealed grease and effluence in the sewers underneath the city; spending time with the nuns of Tyburn who stay up throughout the night to pray for the souls of Londoners; accompanying marine patrol as they scan the Thames for midnight corpses; hanging out with exorcists, graffiti writers and east-London pirate DJs as they play an elaborate game of cat and mouse with the police.

What will I find on these forays across London at night?

I've no idea: that's exactly why I'm going on them.

II

The Panoptic Sublime:

Avian Police

THE PANOPTIC SUBLIME:

AVIAN POLICE

Out on the edges of metroland in Loughton, Essex, along the traditional escape route for East Enders bolting from tuberculoid tenements to roomier suburban homes, lies a strange and little-known place called Lippitt's Hill Camp. Black cab drivers get lost looking for it. When they do, it's hard for them to call for help: the thick foliage of nearby Epping Forest causes mobile phone signals to splutter and die.

The Camp is recessive by nature. It shrinks from publicity. But it has a history to snag the imagination of all those with a taste for the subterranean and covert. During World War Two, and right up until 1948, it was home to hundreds of Germans and Italian prisoners of war. A statue inside its iron gates marks their stay: "Cut out of concrete by Rudi Weber 540177 while POW at this camp October 1946". Later, the Royal Artillery Anti-Aircraft Operations was based here. Underground bunkers and gun pits are still visible.

These days, Lippitt's Hill is used as a base by the Metropolitan Police Air Support Unit. Its officers are conquistadors of the London night. Each evening two crack teams of men and women set off from here and fly by helicopter across the capital. They're edge chasers, hurtling towards those spaces where ground officers fear – or are taught to fear – to tread: rooftops, railway lines, river banks. Their 500k machines, brimming with state-of-the-art hardware, allow them to scan the sprawling city, navigate it at great pace and elevation, and tail anyone who tries to carry out crimes under cover of darkness.

These avian police see, not just a side, but the entire face of London. The rest of us, victims of gravity, stranded down on the ground, have to make do with squinting from the windows of EasyJet planes or going for an evening ride on the London Eye. They do a job many of us can only dream of doing. Many of them, when they started out as beat bobbies, dreamed of taking to the skies too. Sometimes, after they had just been gobbed at by junkie-pimps they were trying to arrest in Camden, or as they stood on Horse Guards Parade in full tunic and helmet with the midday sun turning the backs of their necks scarlet, they would hear the distant roar of overhead helicopters and think how nice it would be to be winched up.

The avian police see themselves as regular officers; before they begin their 7–7 shifts they sit around in the barracks watching European Championship football on Sky or reading authorised biographies of Ronnie Barker. But they are a select bunch: out of 30,000 PCs in the Met, only eighteen of them work in this Unit. As soon as the first call comes in, they leap into action, transformed by circumstances and technology into helmet and harness-sporting redeemers of the night's blackness.

They become Supermen. Their helicopters give them exceptional powers. They can zoom all the way across the congested city in less than ten minutes. High-power lenses and thermal imagers allow them to make out the crocodile logo on a clubber's T-shirt from 2000 feet in the air, read number plates in the dark, look through the windows of Canary Wharf and spot canoodling office workers from eight miles away.

They can fly within fifty feet of Big Ben and cause global reper-cussions by shutting down Heathrow Airport. They can light up the ground beneath them by shining 30-million-candlepower 'nitesun' torches, and help to rescue suicidal young women and befuddled old people who have got lost in Hampstead Heath. Amateur crooks who have caught the odd episode of *The World's Greatest Police Chases* and think they know the principles of heat-imaging hide under trees rather than, as used to happen, lying flat on fields: still they get detected.

"What's the most beautiful thing you've seen at night?"
"Oh, where do you begin? The mist lying in the valleys takes your breath away. The orange glow of the breaking dawn. Or sometimes when there's a full moon you can see its reflection in the Thames..."

The streets of London are made of gold. But only at night time and only from the sky. They lie there, glimmering like a Hatton Garden window display. Jewelled necklaces winking at us. At Piccadilly Circus and along Oxford Street the refracted neon gives them a ruby-red and emerald-green lustre. "Cracking night, Sukhdev," pipes the pilot, but I am too awestruck by the city's beauty to reply. This is the panoptic sublime.

The helicopter flies in orbit. It waggles and tilts. At times it feels as if the pilot has lost control, an outdoor tightrope walker about to fall to earth. The stomach-nausea is accompanied by bursts of landmark glee: there, in the distance, is Wembley Stadium with cranes and machines perched over it like basketball players primed to slamdunk; there the Fabergé rugby ball of the Swiss Re Tower.

Politicians and demographers often assert that London is overloaded, crammed to the hilt, but from the sky it appears far from congested. The concrete jungle is nowhere to be seen. Even the most built-up areas are punctuated by large expanses of dark forest, empty parts of the city's night-canvas. The capital is an endless origami unfolding, stretching out horizontally rather than vertically. Its residential buildings are so crabbed and timid that any sticking out appear both heroic and lonely.

The sky is constantly lit up as private planes carrying Russian billionaires to vital soccer fixtures start their descent, and a whole queue of commercial airliners begin stacking to come into Heathrow; the effect is that of a corporate quasar game as lights continually strafe the darkness. Things invisible at ground level suddenly rear into view: industrial parks – there seem to be hundreds of them. And while, even from the Primrose Hill or Greenwich Park, the city melds into one largely unindividuated flatscape, at night time it becomes more composite in character, a loose and disconnected set of Lego pieces. One pilot describes Croydon as "an oasis of high-rise buildings, sitting there like downtown Dallas."

"Do I like night time?"
"Of course. When I was a bobby I used to go to the top of this tower block that had Big Ben in one direction and Tower Bridge in the other. At 3.30 am, 22 floors up, you could hear it strike: for that few minutes London was quiet and you could hear the birds singing."

The avian police, like black cab drivers, rely on The Knowledge. Spatial awareness is crucial. They can't afford to be disorientated even though they may have had to spin around a trouble spot half a dozen times and at different speeds and altitudes. New moving map systems allow the back-seat navigator to type in any address or postcode into a machine and, at the press of a button on the camera controller, have the camera pointing not just in the approximate vicinity of their destination, but at the precise house number or floor level.

It helps them to be as precise as homing pigeons. Yet, with single Global Positioning System screens able to accommodate only a quarter of an *A-Z* page, and the threat of computer malfunction an ongoing possibility, few officers have thrown away their guide books.

They carry customised cartographies, special versions of the Collins Atlas in which traditional features have been overlaid with police boundaries and London letter-codes for all the police stations. They each supplement this information with biro'd squiggles of their own, creolizing the official geographies with private mnemonics for particular parts of the city. The lights near Wormwood Scrubs resemble a dog bone.

Before the moving maps were introduced, pilots learned to navigate London by its dark areas. Well-known landmarks were often obscured by fog, so they were forced to become airborne sensualists, feeling out and alert to the shapes and contours of forests and verdant areas. Victoria Park, it was generally agreed, looked like a boot.

"What's it like in winter?"
"It can be like being inside one of those snow balls. The snow comes at you from every angle. White dots spitting at you. If you watch *Star Trek*, it's like going into warp drive with all the stars coming at you. You don't experience snow like that normally – travelling at that speed and horizontally."

It's overcast tonight. The clouds we skim and fly through are disorientating. They make it seem as if smoke is rising, as if the city is ablaze. We hover briefly above a mist-obscured St Paul's Cathedral and for a moment I feel I have been transported back to World War Two and the scene of that iconic photograph of Blitz London in which Christopher Wren's dome is surrounded by acres of bloodied devastation.

The helicopter's thermal-imaging cameras irradiate the city. It looks skeletal, postmortemed. On the screen, its nocturnal hues and tints are reduced to black-and-white heat traces. Bleached and decoloured, it has become furtive, like a Customs' X-ray of immigrants smuggling into the country in the back of a lorry. The cameras induce suspicion: why has that snake of light suddenly concealed itself? – actually, it's just a train that's entered a tunnel. Every moving vehicle, at least initially, appears to be a portable terror-container, a nuisance bundle to be monitored and tracked.

The attack on the World Trade Center and now the suicide bombings in London have led to the avian police being placed on constant alert. Heli-routes that fly near key financial and political institutions are almost out of bounds. Each day the pilots are supplied with security updates which assess the threat from terrorists to 'prominent and representational interests' belonging to the USA and Israel. One pilot describes his colleagues below as "ground troops". The thermal imagers themselves, though they're designed to help the police protect the city, produce images that resemble Baghdad, Vietnam – bombing zones for Allied troops. For a moment, London's nocturnal beauty vanishes: the forests seem ash-charred, lit-up areas ghostly apparitions.

"What's the most beautiful thing I've seen? After a storm – when the city looks so washed and lovely. When it's a misty night, you can just make out the tops of high buildings like Canary Wharf: they look like islands in the mist..."

The avian police have to listen to six radio channels at once, a non-stop, mid-air, crosstown traffic of police sirens, command-centre requests, breaking news about pick-axe-wielding Turks on rooftops, random bursts of white noise. But they are also soundscape artists who bring noise to night London, calibrating it to create minimal or maximal impact. Too much roar gives suspects sound cover to break windows or climb fences. Too little, and would-be criminals think they can do as they please.

It would be an exaggeration to call the police sonic terrorists. But they do use sound as a weapon. On quiet nights without remand-centre breakouts or high-speed motorway chases, they fly out to patrol London's crime hotspots in poorer boroughs such as Brent, Tower Hamlets and Southwark. When they see clumps of youths hanging around, they make the equivalent of a handbrake turn in the air. The blades cut the air harder. There is a loud thumping and chopping sound and everyone looks up to see the word 'police' on the underside of their machines. "They've all watched *Air Wolf* and think we can see through walls," laughs one pilot, "We're not going to tell them otherwise."

"What do we see that no one else sees? You can see everyone's swimming pools. There's some absolutely outstanding roof gardens down in the City. Certain areas like Chingford have a love of chequered patios. You've got a run on orange and cream tiles in Barking and East London at the moment: a B&Q lorry must have been turned over."

Flying over a city, especially at night time, allows a brief glimpse of freedom. It is to be liberated from the stress and murk of terrestrial life. Towards the end of their shifts, as darkness slides almost imperceptibly towards dawn, the avian police start to fly back to Lippitt's Hill Camp. Their heads ache and their backs are sore, but though they're at a low-ebb physically, for a few minutes they relax a little and let their minds wander. They think of their families and of past loves. They look at the line of pollution that hangs above the city, so thick they could walk on it, and wish it could be disappeared. They look at the city twitching into motion below them and are touched by its fragility. How beautiful Hampstead looks as it rises out of the mist.

A pilot, his operational lingo replaced by dreamy reverie, reflects on his working life in London:

"When I was working on the ground I certainly didn't like the city. Quite the opposite. But everywhere's lovely from the air.

Even the worst bits look good. Like King's Cross: I never noticed the architecture of St Pancras before – all the stations and the buildings are fantastic. To be honest, I'd rather spend more time in the air than on the ground. Whatever you see on the horizon you can go to. You feel like a giant because the world is smaller."

III

Aborigines and

Unfortunates: Cleaners

ABORIGINES AND UNFORTUNATES:

CLEANERS

The night cleaners of London don't think of themselves as cleaners. They're sure, or at least they fervently hope, that stooping for a living while the rest of the city sleeps is just a temporary phase. The majority are from Africa. Memories of butchered relatives and hazardous exoduses are lodged raw in their minds. But they also harbour lofty ambitions of becoming retail champs and shipping magnates. In their few off-hours they watch CNN and pore over the international finance pages of the broadsheets hoping to glean information that they can use when they return to Africa to set up small import-export businesses. Few will succeed.

The London that they see is a negative universe of public assaults and of swaggering, feral kids. An ungodly realm of out-of-towners on the lash, out-of-control girls spewing obscenities. A mental asylum where the pursuit of idiot pleasures has become, unknown to most of the people who live there, a fatal addiction. They dream of another place, an over-the-rainbow utopia which, more often than not, turns out to be...

"Dubai. It is so lovely in Dubai. I have never myself been. But a friend of mine tells me they have 300-acre ski resorts there. Yes, imagine! And beautiful man-made lights. And the grass is good too. Very green. A military place, you say? Certainly. Perhaps they should have martial law in England. The laws are too soft here."

Each cleaner is an underpaid, under-liveried King Canute trying to push back the tide of over-consumption to which the city is prey. They talk with disbelief at the six tonnes of waste that the West End hotels produce each day of the week. They can rattle down a largely deserted street in their refuse trucks and know, according to which micro-section of the London borough they're in, how overflowing the pavements will be by 1 am or 3 am.

Pleasure, far from being spontaneous and unpredictable, is easily calibrated. The end of each month is the worst time: Londoners are pay-check flush, waving wads of £20 notes or flashing their credit cards, celebrating their temporary liquidity by pissing and upchucking everywhere. The cleaners, present at a party from which they feel estranged, shake their heads at such ritualised abandon. The city's night-life seems to them to be a collective insanity. They see party goers as nocturnal creatures, reckless beasts who slip into the city under cover of darkness to cause mayhem.

Cleaners strive to make early-commute Londoners think that there has been an overnight snowstorm. Every day should be a new day, a tabula rasa rather than a palimpsest. They try to abolish all traces of the previous day. If the city is a text, then cleaners do their best to erase the jottings and doodles that have been inscribed on it.

They operate in the aftermath. After the gold rush. They are instant archaeologists, rapid-response stoopers for syringes, fag ends, gig stubs, demonstration placards. They're also alive to the present and future immiseration of the city, gazing impotently at an anti-spectacle of ragged trolls snouting through bins for half-smoked cigarettes and half-eaten burgers; crazies launching themselves head-first at brick walls; homeless guys clambering into the bottle-recycling skips to sleep.

It's left to them to mop up after suicidees jump from high rises or deranged junkies hurl infant children from balconies. A hardened lot, not prone to sentiment, few can stop themselves holding back tears when they recall the first time they arrived on such a scene and were confronted with dispersed chunks of blood, bones and crushed cloth.

Refuse collectors are exterior designers. Over time they cultivate a keen sense of what is an appropriate beauty for the dishevelled streets they roam. Some weeds they'll let go on the grounds that they give a pleasantly verdant feel to pavements. Coke cans on junction boxes are intolerable though. "We see things in a way that other Londoners don't," one Clapham collector says. "We look at recesses, at the edges of things and under things too. When there's a busy junction, and there are railings to stop people crossing the road, very often you'll find immediately below the railings a build up of dust and detritus with a hard crust on it. It's because a sweeper hasn't attended to it; we have to run a shovel down the side. Sometimes, when I'm on holiday – like when I was in Florida with my wife – I said to her 'Look at the filth on the streets!'

"You need to be able to smile on a nightshift. So people know you wish them well and so that they wish you well. And of course you have a laugh sometimes; like when you see some guy go into pub and emerge four hours later, with a girl on his arm, or tripping and totally bladdered. Or when you see people who have lost their keys shinning up to get into their flats. But for the most part we become the street, the blank architecture. We're there in the same way as a lamp post is there. We're just part of the furniture."

Aborigines. That's what Papa, one of the cleaners at Tottenham Court Road station, calls the tens of thousands of commuters who skelter past him as he sweeps the Underground floors. He suspects they may belong to another civilisation. Racing, frowning, dashing – always in flight to some profoundly important destination. Even the girls with scanty dresses or the mascara-clad boys out to pout at

Nag Nag Nag seem to be in a rush. Their speed makes them, in his eyes, insubstantial. Hollowed men and women. "They are ghosts," he announces, "Dead spirits."

But Papa is no reverse snob. He knows and feels all too acutely the pain of his fellow workers: "We are The Unfortunates." They are students whose money has run out, family men with dodgy visas trying to support their wives and children back in Ghana, unskilled guys trying to make a go of things in the city. They're all too poor to travel to work by tube – the private company that employs them doesn't offer discount tickets – so they arrive on buses. During the winter, it's common for them not to see any daylight at all: they return home from their night shifts at 7.30 am, fall asleep until 4 pm, only to return to work in darkness.

The cleaners can't afford not to be disciplined. They apply method and rigour into getting through the night. The thought of it stretching on endlessly is painful, so in their minds they lay it on a chopping board and slice it into sensible portions, navigable spin cycles of thirty or sixty or ninety minutes. They regard litter not as a sign of the city's opulence or as an assertion of its teeming liveliness, but as evidence of Londoners' lack of focus and proportion. They watch with bemusement and sometimes disgust as young men and, most horrifically in their eyes, young women totter the platforms in a hollering, pissed-up blur. Who, they wonder, are really the lowly ones: us diligents trying to save up for a two-bedroom semi in Southwark, or these cackling short-skirts who cannot even keep their breasts hidden?

This temporary pan-African community clings together for comfort. Its members – from Togo, Nigeria and Ghana – can be found in areas marked 'No Entry', in rooms little bigger than broom cupboards, knocking back one water-dispenser beaker after another to combat the sweltering conditions caused by faulty heating. They listen and add to underground information networks, many of them comprised of gossip masquerading as fact, about fresh passport scams, family-benefit concessions the government has introduced, new contractors who offer cleaning recruits an extra week's holiday each year. They heap good tidings on their colleagues who found a tenner or picked up a mobile phone near one of the tracks. Football, particularly their adoration for Arsenal's Thierry Henry ("He is like an emperor"), also unites them.

Mostly, though, it is a low-simmer sadness that they have in common. Some are getting old, beyond the age when they could imagine another more lustrous future ahead of them. They feel that those few Londoners who notice them presume they are illiterate and not worthy of respect. Night time, they know, is for lying down, not for bending down to pick up other peoples' trash. The past-midnight subway, often as noisy as the African market towns from which they hail, on account of the cross-roar of computer technicians, escalator repairers and track workers, can also fall silent suddenly. And it's then that they begin to hear noises; to spot, fleeing away from them into a distant tunnel, the ghosts of their former selves.

"I think London would collapse if the cleaners would go on strike for just one day. If they were radical the whole of London would be a mess. 'They' means not just the men that clean the Underground, but the streets and the toilets too. Without them you'd see one big mess."

London's cleaners don't exist. Those sleeping take their work for granted. Even those who do see them scuttling across roads in their overalls and starchy, non-flammable uniforms tend to look straight through them. Night time is all about glamour these days, its promise and its most heady realisation. But there's nothing glamorous about cleaners. They may as well be dead. They certainly appear to be only half-alive. In they creak, pushing distractedly at the revolving

doors of the sleek corporate towers where they labour. They're
sweat-glazed from rushing across town. Some have had to cut the
last few minutes of their evening law classes in order to clock on
promptly; others have come from launderette or corner-store jobs;
others have been on the phone for hours desperately trying to get
someone to look after their sick kids for them. They're exhausted
by the time they arrive. By the time they finish, they're utterly spent.

London's cleaners don't exist. Some, employed by violently
penny-pinching sub-sub-contractors, are illegal migrants whose
names are not to be found on any official financial records. They
have no recourse to the law if parts of the salaries are randomly
docked, or if they get hurt because of shoddy safety equipment, or
if they are sexually harassed. So they keep their heads down, their
lips tightly shut. Always, even though they're doing jobs no one else
wants, lifting up to 750 bins on each floor, they feel as if they are
interlopers. Those filing into the HSBC building near Canary
Wharf have their bags checked as they go in and as they leave. Their
movements are tracked and monitored by banks of cameras which
are operated by a control centre in the basement.

Late-working office staff do not look at them though. In
shared lifts, they peer at their feet or suddenly feel an urge to start
Blackberrying colleagues. But night time, even in AC'd corporate
spaces, brings them into unexpected contact with the kinds of civilians their work insulates them from during daytime. They feel tar-
nished, a little afraid, awkward. Some, the cleaners are convinced,
regard them as no better than the rubbish they pick up or hoover.
They rarely smile, or say hello, or seem to have any inkling that the
green-dungareed men and women beside them were once small
businessmen themselves, aspiring politicians chased out of their
home countries by blood-lusting guerrillas, junior-school teachers
who taught orphaned children to read books.

The cleaners themselves do look around, even more slyly than the cameras tailing them. The younger ones comb the open-plan offices for desks under which, much to the annoyance of their supervisors, they can squat and yellow-highlight passages from structural engineering textbooks. Others peer at the photographs that line the walls and show what the building looked like at different stages of the construction process. They wonder: do the CEOs here – those who earn £2680 rather than £5 an hour, those who are driven in by chauffeurs rather than slash-seated public transport, those who have vintage wines and dvds in their offices and who will receive golden handshakes when they leave – do these captains of industry regard us as part of that process? Will anyone commemorate the work we do? We the pensionless ones. We who are not even entitled to sick pay.

And then, sometimes, as dawn is rising, the cleaners take a break from crumb-picking and mousetrap-shifting. Their night's work is almost over. The offices are as clean as the hills and golf courses of the foreign kingdoms to which they dream of migrating. They stand up tall, proud of the reformations they have wrought. Just for a minute or two, they allow themselves the luxury of imagining that they are the shirt-tucked, chauffeur-driven Masters of the Universe who lord it over the snooty pen-pushers and keyboard-dabbers whose garbage they have spent the last seven hours collecting. "Clear out your desks and leave!" they fantasise of declaring.

They wander over to the windows. Light is flooding in, and they feel their spirits rise. They crack a few jokes, whip out their flashy mobile phones ("Hey! Mr Nana from Ghana! Say cheese!"), and take snaps of each other. Then they'll look out over the strange multinational island outside: the helipads; the Millennium Dome; the River Thames speckled with private boats; the top of Canary Wharf. They know it's a republic in which they work, but do not live. They know they are but temporary guests.

Still, for a moment or two, they are struck by the hard, lunar beauty of it all. There, in the distance, is what's left of last night's full moon; it reminds them of nights long ago, thousands of miles away, nights when they kissed their lovers and made solemn promises to always be true, nights when they looked up at and vowed that life would one day be different. They focus their viewfinders and take a photo of a bridge on the horizon. Where does it go? It's a question that nags them all day.

IV

An Atlas of Suffering:

Samaritans

AN ATLAS OF SUFFERING:

SAMARITANS

3am is eternal. 3am is infernal. It's the hour of the wolf. The time at which fear and sadness and regret rack up so that it becomes impossible to get to sleep. Insomnia and self-pity: it's a recipe for hysteria, for wild, lunging desperation. 3am is the dark heart of the city, when the carefully repressed anxieties, aspirations and dreams of its emotionally parched inhabitants can no longer be contained. The silent night amplifies the din in our skulls, returns us to a primal solitude.

There is nothing to be done at 3am except hold on. We stare at ceilings, play old melodies on repeat, curl into foetal balls, stare at old photos, sniff the bed sheets, dial the numbers of people we have not seen or heard from for an age. The pain refuses to go away. We step outside and pace the streets, walk the dog for miles, find a bar to prop up for a few hours, head for a canal pathway where we sit on benches desultorily watching joggers and cottagers go by. Still the pain refuses to go away. We are stuck, impaled between muteness and wanting to scream, madness and cold reason.

In an anonymous office located in a quiet Soho back street, two tired-eyed volunteers are sitting in front of telephones listening intently to people who they have never met before talk about how they are going to kill themselves that night. These volunteers are Samaritans. They have driven or cycled in, sometimes from as far away as Chiswick, to operate what they label the Night Watch. But they watch or patrol nothing. Their terrain is auditory. In a city where friendship is costed, where hundreds of chat-lines and sex-lines price their conversations down to the second, the Samaritans will listen for free to anyone who calls them.

Each call is a journey. It sends them to new social and psychological spaces which they, none of them professionals, must wade through with tact and caution. They have chosen to place themselves on the frontline of the city. They receive its emotional sewage untreated and unfiltered. Minute by minute, hour after hour, they are confronted by the wretchedness from which the codes and civilities of daytime shield them.

The London that they know is an atlas of suffering. They tap into a forcefield of unhappiness and isolation: a lone security guard at an industrial park in Redbridge gazing at CCTV footage while wishing he could be at home tending to his sick child; a housewife, the husband who gave her the black eye she sports asleep upstairs, trying to summon up the courage to move out and start a new life in a new town; a runaway teenager, fresh off the train from Newcastle, wandering the streets of King's Cross while looking in vain for a homeless shelter before resorting to McDonald's where he will try to make a plastic cup of foamy tea last the whole night. The widowers, the orphaned, the smack-heads, the cutters, the bingers, the pre-op trannies, the refugees, the wives of Japanese businessmen: all of them are desperate, all of them are at a dead end.

"There's a question I wanted to ask."
"Of course."
"Well, I had a friend once. He was addicted to prostitutes. He said he had to stop because he found that he had begun to think of every woman that he saw as a prostitute. Or a potential prostitute. I wonder – does your job shape the way you see the people around you? When you're walking around London do

you always see the skull behind the flesh? – Sorry. That probably sounded really offensive."
"The truth is that anyone who works for the Samaritans is already predisposed to seeing the skull behind the flesh."

The Samaritans' centre is an unassuming non-place shorn of character or individuality. Yet, for the needy and the unable, it is an archipelago of hope, a lighthouse in a darkened, miserly city. Its central call room, separated into booths, looks like a language laboratory. It is deliberately neutral, sterile, history-free. The metropolitan stories that emerge into this space do not leak into crevices or mass and hover in the air like a peasouper. The walls do not have ears.

No one likes doing Night Shift. It's the most physically and emotionally draining of all the slots, so much so that volunteers are obliged to do it just once a month. The first part lasts from 10pm to 3am during which at least five volunteers, guided by a supervisor, answer the phones. But from 3am until 8am only two people are left. They will have been sleeping in the tiny, monastic cells upstairs, or flicking through back copies of *Red* in the library; it's hard for them not to feel a little spaced-out or traitors to their body-clocks. They're not health-care experts, but nor, as one of them insists, are they "kaftan-wearing, granola-munching Christians"; they're professionals, mostly white, whose close friends killed themselves at university, who struggled for years to come to terms with their sexuality, who do not find their day jobs boosting the profits of foreign media moguls sufficiently rewarding.

Night Watch volunteers sometimes think of themselves as spies, round-the-clock members of a secret service who not only get to hear the kinds of things under cover of darkness that no one else does, but for reasons of confidentiality cannot tell anyone about those conversations. They manacle themselves to narrative, teasing out and holding onto whatever gobbets of autobiography their callers feel able to reveal. They have to befriend everyone, from those whose droning voices almost put them to sleep, to the Tooting civil servant who abuses his children. But these friendships are fabricated, remote. They rarely tell the callers about themselves and never proffer advice. They become huge sponges, absorbing hurt and perplexity from sprawling suburbs, gated compounds in Kensington, to high-rise estates in Peckham. Sometimes, because even the desperate are likely to see themselves as consumers these days, callers hang up because volunteers may be too young, men rather than women. One volunteer recalls that the first call he ever received lasted two seconds: "Oh, I don't like the sound of your voice."

"Night time is meant to balm or soothe us, isn't it?"
"Yes. I find sometimes that all this negativity corrodes me. It pours into me. Around me. Normally we have boundaries that would allow us to collapse against this excess. So I have to fight to be blasé. To tell myself I have volition to re-experience life in a positive way."

The streets outside the Samaritans' HQ are chocka with lines of muscle Marys trying to get into late-licensed clubs, European teenagers cruising for caffeine fixes, besuited geezers getting out of taxis and heading towards secret gambling joints. These night jays, hungry for excess, craving overload and oblivion, ricochet around as if they're in flashing pinball machines. Are they the people the lonely aspire to be? Or are they, beneath the surface, behind the flesh, also hollowed out?

The manufactured ecstasy of a place like Soho barely exists in the eyes of the lonely. Some, upon seeing it, feel a violent longing to join in. But for most, unhappiness makes London disappear: the black and inky-blue sky, and the neon-lit streets below it, are replaced with unending white space. An Ice-Age architecture.

White is the colour of loneliness; it flattens and razes everything around it. All the buildings, thoroughfares and landscapes by which they negotiate their daily commutes have turned into a vast, aching tundra waste.

Loneliness makes speaking very hard too. The calls that those on Night Watch receive are different from those of their colleagues during the daytime or those who work from 8 pm to 3 am. They are less likely to come from cranks or sex pests or insomniacs simply wanting a chat. They tend to come from a deeper, darker space than those, sometimes made with mobile phones, from young adults who have just been dumped at a Camden pub by their boyfriends. The very lateness of the calls gives them a different texture and gravity, as if the stories narrated in them are inflected by the woozy, soporific atmosphere into which they tentatively emerge. Even those who are ringing from their own flats tend to speak more quietly than normal, in tribute perhaps to the fact that night time is for whispers as much as it is for cries. The effect is to create a complicity with their interlocutors.

"We are part of a space like no other," reflects one Samaritan. "It's a space of incredible intimacy, of anonymous revelation." Often though, callers stay on the line for over a quarter of an hour without saying a word. But the silence is not silent. Ambient sounds – from dampened sobs to background radios – can be made out. More than that, the Samaritans learn to gauge the quality and consistency of that silence, to appreciate its rhythms and contours, knowing that to break it too soon would be an act of violence, but punctuating it occasionally with a gentle, "We're still here" or "Take your time."

Sometimes, the silence is purposeful. It has a mute dynamism. The Samaritans feel sure that if they can just hold out a little longer then a volley of revelations, whether in gaspy fragments or violent outpourings, will likely follow. These trauma narratives are often delivered in serial form over the course of many nights or even weeks. They may be sustained confessions, but the Samaritans are not there to forgive or deliver redemption. They are mere funnels. And yet, the calls can make an impact upon them, invading their dreams and getting beneath their skin, revealing doppelgänger individuals, parallel characters whose lives by pure chance have jagged in an unfortunate direction. They don't wear headless microphones which they think would transform them into tele-workers or call-centre operatives; as a result, long calls leave their eardrums hurting, their ears red. The conversations etch themselves onto the Samaritans' flesh.

"Why do you do this?"
"It gets harder and harder to remember. Sometimes you wonder if you're a voyeur. If the calls are a way of getting a fix."
"How is that?"
"You have to ask yourself what you're getting out of the call. The exchange can act as a poultice, as a way to reaffirm to yourself that things are going well in your life, both because you're helping and – this is the fetishistic element – your life is not like theirs. It's a dangerous feeling. But you'd be lying to deny that it's part of the palate of emotions you go through."

The night inches forward. For the lonely it's a marathon whose wall they hit long ago. They exist in a time zone of their own, barely aware when they ring the Samaritans that it's such a late hour. 3 o'clock. 4 o'clock. 5 o'clock. In winter, dawn is barely distinguishable from darkness. There's scarcely any light at the end of the tunnel. Their calls may go on and on. The Samaritans are the only safety blanket they have; their conversations a long-distance embrace.

For the volunteers themselves the absurdity mounts as the hours tick by. They try to relax in between calls, playing chess with their partner until the phone rings once more. But it's hard for them not to feel like vomitoria into which are being hurled all of London's rejection, pathos and abjection. Everyone they speak to is tottering, precarious. And this city, this imperial centre, this bastion of government and power and fortune-making: suddenly it seems liquid and unreal too. It's full of millions of people, half-persons, unmoored from the families or solidarities that might give them ballast and belonging, always swimming against the tide, always a breath away from submersion. How, they wonder, can two unshaven, bleary individuals do anything to help? And then, they quickly remind themselves, how can they possibly not try to?

Eventually, the calls wind down. Night Watch is over. The volunteers go upstairs for a shower and a shave. Some get on their bicycles and start pedalling home. They feel knackered, jetlagged. But also the uplift of survival, of getting through another long night unharmed. As they cycle back to Hampstead or to Putney, they breathe in the petrol-soaked air. Suddenly it smells beautiful to them. They pass by building sites, hear motorists yell curses at them. The clamour is ugly and affirming.

They move through areas of blight and poverty. If they were normal commuters, they would probably fiddle with the reception for Kiss FM, or make a quick mobile phone-call to their work colleagues. But to them these cartographies bleed biographies. They peer at the council estates, acutely aware that it was from places like these that some of their callers last night were ringing.

They pass long, straggling bus queues and observe the commuters heading for the Underground. They seem, on the surface, so forceful and in control. But to the Night Watch volunteers their black business suits and briefcases look all too funereal. They think to themselves: why are they heading back to the jobs and to the places that they say make them feel so sick and unhappy?

"Sometimes, after a really heavy night, when you've just heard tragedy after tragedy, people at the end of their tether and wanting to kill themselves, people struggling because they're so poor, you go out and you see a red bus go by with one of those Mayor of London posters: 'Seven million Londoners, One London'. And you think: '*Really*?'"

V

The War Against Terrors:

The Exorcist

THE WAR AGAINST TERRORS:

THE EXORCIST

Every night London is placed under siege. Every night, across the city, thousands of people are attacked and maimed. The perpe-trators will be ghosts. The victims will be poor, foreign, fragile. Unsurprisingly so; ghosts feed off isolation: few people see them while emailing or telephoning friends. Between midnight, and espe-cially between 2.30 and 3.30, hours when the human body is at its lowest ebb, and when sorrows gnaw most viciously at the heart: that is when they are spotted. Those who see them, or who start to suspect that the tantrums and moodswings of their teenage daugh-ters are signs of demonic possession, are too scared to tell anyone. They live in fear of Social Services knocking down their front doors and taking away their children.

So they stay silent, going slowly out of their minds with distress and confusion. Some, to their later regret, hand over what scant savings they have to sharp-talking Liverpudlians or regal-looking Africans wearing colourful robes. Others end up on the doorsteps of local churches. There then begins a series of sob-strewn and whispered conversations; rabbis, Imams and priests, all of whom would be loathe to go on record to admit the existence of diabolical spirits, open their private address books and, as often as not, tele-phone Norman Palmer to tell him that their flock-members have been given special dispensation to be exorcised.

Norman is a protégé of occult writer Dennis Wheatley and an ex-Marine-turned-gift-shop-owner-turned-paranormal investigator. He is 67 years old, wears gold on his fingers and shell-suit bottoms, and when I meet him is chomping on a cigar despite having consulted his chemotherapist a few hours earlier about the treatment he needs to scrape the cancer off his lungs. He seems like the oldest blinger in town. For over thirty years though he has spent much of his spare time developing a reputation as one of London's leading exorcists. A semi-spectral figure, one whose (unpaid) work elicits fear and sniggers, he's not to be found in the Yellow Pages. "I'm a last resort," he says, "a catalyst in the constant fight between good and evil."

"During the War I'd stand at the bedroom window and see these brilliant searchlights lighting up the sky above Lincoln Inn Fields. You could see the planes trapped in the spotlights. It was like a Marvel comic. The dogfights were eerie because of their soundlessness. You couldn't actually hear the screaming engines or the machine guns. It was strange, like a silent film. Every once in a while you'd see a plane hit and start smoking. You might even see the pilots bail out or floating down in parachutes over London. But you didn't know exactly where they'd come from or where they'd landed. They'd just disappear."

Norman is part marksman, part waste-disposal expert. He seeks to disinfect the city – its hapless inhabitants and haunted spaces. "Two-up two-down semis: that's where the battle against evil is fought out on the ground," he says. "You don't go to the London Dungeons or the Tower of London; you go to wherever Mr and Mrs Bloggs live." He and his wife Yvonne spend much of their spare time walking the side streets and back alleys of London looking for signs of 'messiness'. Bad smells, though not so bad they would be apparent to anyone but the keenest-nosed mutt, are a giveaway.

Ghosts belong to the capital's grime economy: they give off an odour of decay, of offal. Lost souls may smell reasonably fragrant, but evil ones are 'putrid'.

As we enter a late Georgian square in Islington, Yvonne starts to sniff. "Something's not right here. It's not nice. That square's not welcoming. I don't smell sulphur; I smell sewage. It's revolting. Can you feel it?" she asks Norman. "No," he replies. A few minutes later, I find her clutching some railings with her eyes squeezed shut. "I'm feeling coldness. It's not nice," she complains. "I don't like it. I just don't like it." To a disbeliever, the couple's olfactory- and heat-seeking methods of research might seem like little more than outdoors feng shui; many of the buildings they suspect of harbouring bad presences are found on street corners and so are prone to cross-current breezes, while others merely look a little rundown and in need of new net curtains. Yet it's hard not to be struck by the pair's sensualist approach to navigating London. They feel the city rather than read about it; to them its history is not a textual or archival entity, but something fleshy and living.

And dying too. For Norman and Yvonne feel increasingly estranged by the changes in their neighbourhood. As they wander the deserted streets of Clerkenwell they find themselves preoccupied at least as much by what used to be there as much as what developers have allowed to remain standing. The pavements are full of the ghosts of friends and neighbours, most of whom have been priced out by recent gentrification. Occasionally a faded hairdresser or greengrocer sign can be spotted. Mostly, though, everything seems to be in transition: greasy-spoon caffs of the kind they go to each afternoon are disappearing and being replaced by bistros and fancy restaurants. Norman and Yvonne are planning their own departure: to Plymouth.

They will leave behind them a city congested with ghosts, spirits and immaterial forces. Theologians shy away from talking about these presences, but the Palmers, who have decontaminated council-estate flats and financial-zone offices alike, help everyone from recent African migrants to yuppies and beery students whose post-closing-time Ouija board experiments have gone disastrously awry, are as busy as they have ever been. London is over-lit, its streets are monitored by CCTV and the avian police, its inhabitants monitor themselves using webcams, digicams and mobile-phone cameras; yet the nocturnal city can never be wholly regulated. Ghosts represent residual energies, the unruly subconscious, a wild republic fashioned by the human imagination and hence impervious to positivist or sociological assault.

"I joined the Marines in 1956. Thought I was invincible. It was at the time of Suez. Britain was a troublemaker all through the world. There was always some sort of punch up somewhere: Malaya, Burma, Cyprus, Aden. They said what we were doing was strategic withdrawal; it was bloody mass evacuation. You see extraordinary things: I saw my mate, my bosom mate, literally cut in half by a machine gun right next to me. I must have killed about a dozen men. Four of them close up. You feel almost callous, but it's self-preservation. You have to put that wall up between you and those you're fighting. Then in the 1960s I became a mercenary, getting involved in covert operations against enemies that Britain wasn't meant to be officially fighting...."

"You'll notice I'm not wearing a cloak and pointy hat," says Norman. It's a cold Friday evening in autumn and we're here to conduct an exorcism of what he declares to be "the wickedest place in London." The House of Detention (motto: 'Look on and despair all ye who enter here') is 20,000 square feet of underground passages, tunnels and cellars that lies in a secluded part of Clerkenwell. It was established in 1845 as a holding joint for sheep stealers, loaf thieves, cutpurses and serial killers. It was a remand prison, an indeterminate zone where the poor, wild and luckless lingered a while before being packed off to other gaols or, in some cases, to another world: Australia. Large portions of it were subterranean, making it a hard place from which to escape. At its peak, or perhaps its deepest trough, 10,000 boys and men were squeezed within its vice-like grip. It was crowded as hell.

By the middle of the 1860s it had become a radical pen caging growing numbers of Fenians; in 1867 a botched escape plan involving explosives killed twelve prisoners and local people, and wounded another 120 of them. The jail was closed in 1890, but trace memories of the deaths and desperation it housed have persisted to the present day. The residents of an adjacent sheltered home which we're about to enter have been complaining about strange noises and spirits. Some have learned that their building lies on the site of a former nun's cemetery. The unease that this compounds is shared by the home's warden; he says that his relatives refuse to sleep in the upstairs bedroom of his residence because of its creepy atmosphere. Even Norman is nervous: "All week I've been feeling distinctly apprehensive about this. Mark my words, this is a big one."

In fact, the whole atmosphere this Friday night is discomforting. To one side of the warden's flat is St James's Church and on its grounds local kids are boo-yahing, sex-teasing each other and throwing bottles of cheap beer around. In the past they climbed into the warden's garden to retrieve footballs they'd kicked over, but to stop this from happening he has slung a ten-foot-high net above it as well as placing spikes on the wall. This produces a feeling of incarceration rather than comfort, a feeling that's shared by some of the older residents (at one point he calls them "inmates") who live in the complex. The communal garden is full of benches the widows and widowers have dedicated to their loved ones: 'Dearest Ebby, Your space will never be filled – Husband Tom'.

According to Norman, the House of Detention and the gaol which it replaced in the 19th century have incubated and built up layers of unimaginable wickedness: "Think of all that sorrow and fear and anger and negative rage within those four walls." Because the exorcism is being conducted remotely, within the radius of the infection rather than at the source of the infection, he's worried that he will have to expend far more energy than he had anticipated. While he mops his brow and asks the warden to identify recent signs, Yvonne and I go upstairs to scout the building.

The room in which the warden's relatives refuse to sleep is indeed rather creepy. Mainly because its former occupant appears to be a lover of real-life crime and detective stories. The walls are lined with original covers of *The Illustrated Police News*: "When Will The Whitechapel Murderer Be Captured?" Shelves are full of books and videos about Jack The Ripper, Richard Nixon, and Oscar Wilde's imprisonment. Yvonne, thermally rather than textually inclined, says she's confused. "Here: I do NOT like it. It's really really creepy. I'm not cold, but my hair's standing on end. I can feel the draught coming from there." She points to a spot near where I'm standing.

Am I myself a ghost? Suddenly, in this strange house, with noisy London and all its up-for-it clubbers and Friday-night ravers apparently a million miles away from where we are, anything seems possible. It's only when we return downstairs that Norman declares the source of the problem is not coming from the residence itself, but is actually part of the malign forcefield exerted by the House of Detention: "The bad news," he tells Yvonne, "is that it's the old enemy we've come across before. I'll give you three guesses where." "Guildford?" "Yes. It's Thentus." Thentus, it turns out, is a high-ranking demon, a manifestation of pure evil, one that they thought they had banished during a previous exorcism in suburban Surrey. Evil is an elusive migrant too wily and centrifugal ever to be totally extinguished.

And so, with the living room lights on and the curtains undrawn, Norman sits down on an armchair and begins to take out paraphernalia from his plastic bag. Apparently even those spirits attacking Hindus or Jews respond most effectively, that is fearfully, to Christian weapons of destruction: a Bible, a crucifix, holy water that he has brought along in a small plastic bottle bearing a sticky label with the words 'Holy water' written on it. Those spirits may be old ones, they may hail from the sixteenth century, but they always have a perfect understanding of contemporary English.

Norman starts to recite the Lord's Prayer. He speaks with such feeling and pauses so lengthily between lines that I worry he's going to keel over. His breathing is erratic. I know he's in pain because his lung cancer requires him to take constant pain killers and medication, but he has left his tablets at home. He's not allowed any alcohol which might help mask the agony. Yvonne urges him not to sit back on his chair as it might exacerbate his piles. "My head is tight," he complains. "I'm getting a splitting headache."

On the walls of the room hang nineteenth-century prints of London scenes: John O'Connor's *Ludgate Evening* and Charles Dixon's *Tower Bridge*. But although Norman earlier crossed my forehead and sprinkled me with holy water, there is little art or spectacle on show here. He sees himself as a technician rather than as an artist or a performer. On he labours: "Cease to trouble this place," he commands. Then: "Let evil spirits be put to flight. Deliver this place from the assaults and temptations of the evil one. Before your presence, oh Lord, the arms of hell are put to flight. Free these premises from every evil and unclean spirit that may be assailing it. Begone Satan and cease to trouble this place."

Soon Satan does get himself begone. Or so Norman tells me the following week when we meet to discuss the evening. The exorcism had been expected to last all night long; in the end, it was done and dusted in a couple of hours. It was a very undramatic victory. No hurled crockery. No blood on the walls. No dancing ectoplasm. The strangest thing, as I discovered later in the evening, was the disappearance of my notebook and my mobile phone. "To be honest, I think whatever was there beat a tactical retreat. I think it's trying to lull us into a sense of false security. I'd have been happier if I'd come out of there like a wet dishcloth."

It seems that Thentus, just like Norman and Yvonne, had finally decided to leave Clerkenwell. He had been in the neighbourhood for a good while. Where he has decamped to is unclear, but then the fight against evil, as Norman had often reminded me, is a ceaseless one. Somewhere else in the capital will have to bask in the anti-glory of being dubbed The Wickedest Place in London. But at least the House of Detention is now ghost-free. The residents of the sheltered home opposite it seem happy enough.

Norman, who goes to great lengths to describe himself as less a magus, and more a simple operative who sometimes does a spot of night-shift work, is feeling a little hardier. And thoughtful: "I think everyone's searching for something that's missing. But it makes you vulnerable if you're looking for something unseen or intangible. Things can easily get out of control. I suppose I deal with the substrata of London. Its substructure. You could say this is a parallel London that the real or physical London doesn't know about. It occupies a different plane of existence."

VI

A Partial Burial:

Flushers

A PARTIAL BURIAL:

FLUSHERS

It begins, as nights in London often do, with a rumour. Or rather: a grim promise. I had been told by a friend of a friend about giant hairballs clogging the sewers under the streets of the capital. Over the course of many years, tiny strands of hair moulted by millions of commuters had built up so that, lacquered by grease and dirt, they formed huge furry, knotted boulders that were swelling all the time. These hairballs, my informant whispered, had begun to develop a life force of their own; soon they would be bursting out of the sewers, rolling onto the streets and flattening pedestrians. I wanted to witness them at first hand: I would have to fight them, just as Raquel Welch in *One Million Years BC* fended off murderous brontosauri with a pointy stick.

And so I head to Victoria Embankment where, on a sodden summer's night, I am poised above an open manhole cover on the brink of plunging underground in order, I hope, to face-off against the subterranean monsters that lurk below the city. If they exist, they are fewer in number than they used to be; a few yards away lies a monument to Joseph Bazalgette, the civil engineer who designed London's sewer system, bearing the inscription: "Flumini Vincula Posuit" ("He Chained the River"). He looks as whiskery and fervid as any 19th-century empire builder, and that's what he was: a man driven by the urge to fight choleric murk and miasmic waste, to civilize and pacify the capital.

Here, near the banks of the river, everything is illuminated: the floating restaurants moored a few yards away; a blue-neon-clad Hungerford Bridge; theatres emblazoned with the names of American actors in town to sprinkle a bit of Hollywood stardust on trad productions of repertory staples; St Paul's Cathedral in the distance. Packs of office girls totter home arm in arm after all-evening jolly ups, while lairy joyriders blow their horns and waggle their tongues furiously as they drive past them. Everyone seems hell-bent on having a good time. Everything seems safe and suburban.

To go underground though is to enter harsher, less secure territory. The flushers – or cleaners – who do so have to wear special protective outfits: a white paper boiler suit of the kind once sported by goggle-eyed ravers at German techno clubs; thigh-high waders with metal-toecapped boots; knee-high white socks; a safety helmet with a light attached to it; a steel box containing emergency breathing apparatus; a harness to winch them up in case of sudden floods or respiratory problems. They are kitted out thoroughly enough to deal with the end of the world.

And those who spend their working lives below London do inhabit the end of the world. The bowels of the earth, in fact. A place where day and night are interchangeable. Where darkness, uniquely for the capital, is perpetual. Its very essence. These Styx-dwellers manhandle everything that its inhabitants have abandoned and forgotten. They wade and crawl through its waste – its condoms, nappies, cotton buds, shit, fat – to clear blockages and keep the city above functional. They shuffle and tread through a rat-plagued and endlessly proliferating sub-universe, full of alleys, pipes, tunnels and side-tunnels so numerous that no single map exists to plot them all, that is invisible to most city-dwellers, and would repel them if it were not. If the 100-degree heat was not so intense, and their backs were able to withstand the pain-spasms, and they did not get lost in this sign-less republic, they would be able to trek east all the way from Hampstead to Beckton.

"If you really thought about where you were going and what you were doing you'd either be shit scared or you wouldn't go there. We're shit shovellers. Some of the jobs I do a high percentage of the country would turn around and say: 'Poke that up yer arse mate as far as you can put it.'"

The modern city lionizes the idea of the underground. It is a synonym for buzz, edge, cool – perfumed and dematerialised concepts whose development and manufacture sustain huge swathes of the post-industrial economy. Now that 'downtown' has been gazumped by real-estate moguls, 'below town' has become the new avant-frontier. Paris, in particular, has a legion of catacombists and spelunkers who fashion crepuscular communities below the arrondisements. Yet London, a low-rise city much of whose mythology emanates from the ancient history embedded in its substrata, has no such tradition. The sewers are smaller and far less savoury than those on the Continent or in New York. Only the most daring or stupid adventurer would clamber down a steel staircase to reach them.

The flushers represent a small and dwindling tribe, all of them men, who deal with the reality rather than the poetics of subterranean London. A couple of decades ago there were over three hundred of them; now there are only 39. Their profession has been privatised and contractors, a growing number of them from Eastern Europe, perform many of the functions they used to. Semantic creep means that flushers these days are often labelled 'water technicians'. Most of them have been working underground for two or three decades and this longevity, as well as their relative old age, makes them unusual, picturesque even, in a city where manual work is seen as marginal. "Us poor buggers are treated like bloody pit ponies," one complains, but that sense of injustice (the starting salary is little more than £18,000) is also allied to a fear that theirs is a waning vocation: "Young people today aren't willing to get out of bed every day, especially for the kind of money we earn, just to dive back underground again."

"Have I seen ghosts? Well, you can imagine how that'd go down at the canteen, can't you – 'Time you went home, my son, took a pill and went to bed.' Flushers'll say that they felt a change of draught or had a cold feeling. The mind is a horrible thing. To wander about in the sewers on your own would be a very unwise thing to do. Being underground is like a partial burial, innit?"

The sewers are often imagined as an unpatrolled, deregulated wilderness. They are imagined – and correspondingly desired – as the mephitic opposite of tamed civil society. They are meant to cancel it out. In truth, they mirror rather than reverse what goes on above ground. Flushers who work below Charing Cross Hospital complain about the overpowering smell of ether and about the number of times they have been stabbed by syringes while crawling on their hands and knees. Those who work below garages find themselves up to their necks in petrol. And those who happen to be rearranging the sludge below restaurants swear that they can tell whether it's come from an Indian or a Chinese takeaway.

Fat is the bane of flushers' lives. Millions of litres, from half-eaten breakfast dishes, chip-laden frying pans or fast-food joints, are tipped down into sinks each day. Eventually they find their way into the sewers. They represent the effluence of affluence. They are the graffiti that the contemporary leisurepolis scrawls on subterranean environments. Thirty years ago the Thames, unloved and abandoned, created few problems for flushers; now, the river's banks are congested with clubs, boozy eateries and art-complex gallery cafes all of them disgorging fat.

I wade through some of it at Victoria Embankment. It is at once crunchy and spongy, like putrid bran. Brown and white and grey: a pigeon-shit potage sprinkled with an extra top layer of mop heads and tampons. Flushers tell stories of accidentally getting a gobful of the sewer flies who feed on the fat or of metal grating giving way so that they fell into eight-feet deep fat-quicksands; the mouthfuls of the stuff they swallow leave their guts raw and hollering for months on end. But it's the bouquet that makes their flesh crawl: "You smell it initially. You breathe it all day long. You pass wind and what comes out is the smell of the fat. You can go home and shower as much you like – even with washing-up liquid – but at the end of the day you're still farting the smell of rancid fat. My wife'll say: "Oh, I see you've been sorting Fat Problems out..."

There is one story that many flushers in London like to recount. It concerns a fat iceberg that had been building up below Leicester Square over the course of a whole decade. Eventually, this 150-square-feet "slug of hardened fat" grew so large that it was impassable. A gang of flushers armed with supersucker machines spent six weeks one blazing summer trying to dislodge it. By the time they finished they were reduced to using ice picks to hack away at the white mountain.

"It looks like a huge packet of lard. It shines in the dark and gives off this phenomenal transparent heat. Within ten minutes, as soon as you stick a shovel in it, you could slide through. The water comes at you like a dyke. The risks are colossal. Later, an animal food company got in touch because they wanted to buy and recirculate the fat."

The darkness of the sewers has the effect of amplifying and intensifying sounds. Ears need to work as hard as eyes to aid navigation through the curving tunnels that create an echolalia at once consoling (any cries for help will reverberate for a long time) and terrifying (scary, inexplicable noises will reverberate for a long time). Thin skeins of tree root thread their way down tunnel roofs and walls: they look like hanging microphones that curious city dwellers have unfurled to eavesdrop on a world they cannot see. No traffic or commuter roar can be heard here; rather, they will pick up a constant susurration of distant waterfalls and squelching waders that can be heard alongside the murmur of near-adjacent Underground trains and the electronic ping of a machine that monitors carbon monoxide levels. Perhaps, like the flushers themselves when they stop for a while to adjust their equipment or to take a breath, they will make out the scratching of angry red cockroaches or of the rats that consider the sewers their home.

And, once in a while, they will hear the roar of flushers laughing. They'll cackle as a hungry gang member finds an orange among the dirt and fat and promptly starts eating it. Or at the desperate worker who loosens his uniform and has a dump in a corner. Life in the sewers is hard, and humour – the coarser and blacker the better – raises flagging spirits. A flusher tells me that he bought his wife a bouquet of flowers on their wedding anniversary. "I suppose you'll be expecting me to open my legs for you," she remarked. "A vase'll do," he replied. Another remembers the night he emerged from a sewer at Leicester Square dripping of filth and shit only to find a young woman tourist peering at him. He held out his hand: "Smell that. That's Canal No 5, that is."

"One day, my son, if you work hard, and study all your books, you could get a job like this. Fucking hell, we do a unique job, but we're not designed to go underground: if we were we'd be moles and voles and rats and we'd have super-duper noses, have whiskers on, and be able to dig. It's an alien place for us."

I never did find the giant hairballs I was looking for. Apparently they don't exist. I shouldn't have been surprised: the sewers of London accumulate myths as much as they do fat. They are built out of a sediment of gossip, whisper, untruth, longing. Subterranea is alluring because it is thought, by autodidact dreamers and unstable visionaries, to contain the solution to the city, to store its hidden wiring. The streets become a horizontal curtain below which there paces up and down a magic controller, or at least a magic code that might be deciphered. At a time when gentrification is eroding much of the texture and historical gristle that made London what it was, is and should be, the sewers encourage necessary speculations about the existence of secret tunnels, through time as well as space, that might lead civilians to a parallel metropolis – Arnos Grove-meets-Atlantis, Snaresbrook-on-Shangri-La – in which a richer, deeper urbanism flourishes.

Although the sewers are no paradise they do contain treasures. Inching through them with only a helmet light to pierce the enveloping darkness, distant memories of being captivated by stories about Aladdin's Cave or Howard Carter discovering the tomb of Tutankhamen get rekindled. All flushers strike lucky once in a while and, most commonly under Cricklewood, St John's Wood and Formosa Street in Paddington, happen across something glimmering among the shit: chains, diamond rings, gold sovereigns. Their long-suffering wives are especially delighted.

The sewers are more than negation or black absence. They are rich in fragile beauty. The brickwork, some of it over two hundred years old, amazes all who work below ground. The red bricks and Portland cement, so smooth and enduring, put the modern fashion for concrete to shame. "The joints, the construction – it's marvellous," enthuses a flusher. "The people who built it could have thought: who gives a shit about this? It'll be hidden to most Londoners. It's only a sewer. But it's our workplace."

The light astounds: it bounces off the muddy water and the greasy streams and the curved tunnels at different angles, creating a fly-flecked and dust-particled vision that allows, for one brief but unforgettable moment, the smell and heat to be forgotten. The walls too can be minimalist canvasses, industrial cave drawings that reveal a fading archive of metropolitan graft: dates, some stretching back eighty years, of when repairs were carried out as well as the initials of the repairers. The flushers, unlike Bazalgette, will never be commemorated: some writing their names on underground walls with high-pressure jetters. Over time the endless wash of sewage rubs and distresses these inscriptions and self-memorials – as it does the flushers' bodies. Somehow, just, they endure.

PHOENIX
CARS

VII

Strange Antennae:

Mini-Cab Drivers

MINI
CABS

STRANGE ANTENNAE:

MINI-CAB DRIVERS

"The Georgian night was sustained by port," declared H. V. Morton back in 1926. "The Victorian by champagne: this is the age of Clicquot." These days, nights in London are sustained by vodka and Red Bull. By reheated chunks of battered fish and greasy slabs of ketchup-laden shish kebab. Who says so? The capital's mini-cab drivers; they're the ones who are forced to tissue and hose down the retched-up contents of those meals from their back seats. Like every motorist in London, each one is happy to rehearse a litany of complaints: speed bumps, congestion charges, the drug dealers whose white powder turns the West End into a winter wonderland every night. However, their biggest gripe is always against the teenagers and kidults who vomit and kebabify their cars.

Mini-cab drivers may hate many aspects of London. But they could never hate it as much as London appears to hate them. Passengers yell and curse at them. The Mayor's Transport for London office pastes the city's billboards with posters portraying them as rapists. Swashbuckling blockbusters at cinemas from Piccadilly to Staines are preceded with public information films that depict them as predators of the night, taxi terrorists. Then there are the black-cab drivers: the self-designated custodians of the city's soul blame mini-cab operators for undercutting prices, careering through streets haphazardly, and for opting out of the strict metropolitan disciplinary codes to which they themselves dedicated years of their lives.

Mini cabs, and there are 45,000 licensed ones, may lack the regal, authoritative stature of black cabs, but it is they who sustain the contemporary London night. Without them, clubbers, party-goers and shift-workers would rely wholly on an erratic and seemingly non-existent night-bus system. Their company offices are often found in cheaply-rented basements or at the top of flights of wilting lino'd stairs. The smoke coming from the waiting drivers can be pea-souper-dense and gives rise to the impression that these are dark dens, kindred mal-spirits of the language schools, pawn shops and backstreet massage parlours that sometimes surround them.

They are staffed, as is so much of the night, by recent immigrants to the capital. Many are from Afghanistan, Pakistan and Ghana. They tend to live close to their offices, unlike self-employed black cabbies who are more likely to drive in each evening from Bexleyheath in Kent or, increasingly, fly in every other week from ex-pat villas in Spain. Their fares, bartered by office-controllers rather than metered, are less, their hours longer. Their working conditions veer on the dismal. And yet, perhaps because of their wavering accents, or the general pall of disrepute that hangs over them, and despite recent regulations that means they must fill out reams of official paperwork to prove they're not illegals, they are dismissed as qat chewers, stubbly lechers, illicit sharks, conmen who fiddle the companies for whom they work as well as their hapless passengers.

They will never appear in tourist brochures or in cameo parts in trans-Atlantic romantic comedies. Their social critiques, hopes and dreams will never be echoed or celebrated by newspaper columnists.

"We're faceless and anonymous. To most people who sit at the back of the cab we're just part of a system. In their eyes we're just robots. We have lots of time to think, but they don't think we could engage them in any kind of real, intellectual conversation. They dismiss us and weigh us up: they think we belong to council estates or are abusive to our wives and kids when we're

not driving. We're just an ATM to most people: they shove money at us and don't even say goodnight."

Night cabbies claim they prefer to drive at night because the streets are quieter then and there is less of the pollution and pressure that has sent their colleagues to early graves. Others – laughingly, grudgingly – confess to being misfits. Their cars are secret caves into which they retreat to close out the disappointments and regrets of their daily lives. Night-riding allows them to live in London incognito, to lurk in the shadows away from the glare of family and friends. They stay on the move to keep themselves from thinking, trying to substitute street names for the personal calamities the recollection of which stabs them into bitter consciousness just at that point in the early morning when they are beginning to doze off.

And so the cabs become floating coffins carrying the corpses of countless hopes and future-dreams. Their drivers miss their wives whom they left behind when they got aboard pick-up trucks on the first leg of their long, mountainous journeys to the United Kingdom. They tear up when they think back to the crackling phone call telling them that their youngest child had died in an accident back home. They curse their naivety in assuming that it would be easy for them to upgrade their medical qualifications and get jobs in the NHS.

But even for those cabbies who were born in London, night time triggers bitter-sweet memories that pierce the modernity all around them. The neighbourhoods in which they were raised and through which they now drift are full of phantom architectures, bogus street fronts. The cabbies' minds wander, and they recall how the internet cafes and tech-house clubs beyond the traffic lights on the left were once the dancehalls where they themselves first asked out their future wives, the picture palaces where they stared at and dreamt of being Tony Curtis, the tailor shops where they were measured up for their very first work suit. The places where they became men, where they grew up to become fully-fledged Londoners: all these have disappeared.

The customers too can rekindle long-suppressed memories. A Bangladeshi driver, now in his forties, speaks of the night that he took a passenger, whose wife had just given birth at the Royal London Hospital on Whitechapel Road, back home to Forest Gate. The driver kept glancing at his rear-view mirror, convinced that he had seen the pock-faced man in the back seat before. Finally, just before he pulled up at the passenger's cul-de-sac, he remembered: it was the same guy who, with a bunch of braying mates, had beaten him up for being a Paki in a Stepney back street in the 1970s.

"There's a difference between what people see when they look in and what I look out on. The front window is a film camera. I'm in a self-contained capsule looking out. I'm almost like an alien that moves through the city. Information transmits itself to me. It's as if I have strange antennae. With them I capture people, buildings, sometimes even time itself."

Cabbies regard their rear-view mirrors as microscopes. They use them to peer at their passengers with the expertise of lab technicians. One driver describes the people he picks up as "multi-cellular organisms superimposed on an illusory background." All drivers are part-phrenologists, part-behavioural scientists; they are able, or so they claim, to read off passengers' biographies from their faces, voices, and their body language. "You know what they're going to say before they've even said it," claims another front-seat clairvoyant.

At night time, even if they are carrying magazines or email print-outs they could read, customers almost always ask for the lights in the back seat to be turned off. They like to observe the city from a position of enabling darkness, but they don't want the city to be able to observe them. Suspecting that light has the power to reveal and magnify parts of themselves they would prefer to keep hidden, they baulk at the idea of their souls being spotlighted, of truth being exposed. The darkness allows them to masquerade and to embellish their personalities: on-the-make wide boys fashion themselves as nascent IT millionaires, downtrodden secretaries reinvent themselves as would-be salsa dancers on the verge of marrying Barcelonan boyfriends ten years younger.

Boredom breeds elaborate fictions. As the night wears on, the cab is transformed into a small theatre. The drivers, in spite of complaining that their passengers are truth-twisters and vainglorious grandstanders, become actors themselves. They reveal, with varying degrees of prodding, partial narratives about the countries from which they hail and the apparently bucolic lives of the people who live there. Yes, of course the music and food is great there. Yes, of course they will return. They become salesman for their nations, offering fantastic back stories for the juju and highlife players whose bootlegged tunes they belt out on their stereos. They invent dramatic characters they have driven to penthouses and claim to be the favourite chauffeurs of famous actors. "Mr De Niro tell me to call him up if ever I am in America."

"Londoners are just like rats. They climb over one another. You see the way they try to get to a certain destination, their meanness, the way they fight with each other, try to get to the top of the pile. What is their goal? It is to buy a very expensive car or to buy a swimming pool in their garden. Maybe they want to go on the most expensive holiday they can afford. It is all so blind. There is no morality to them."

Mini-cab drivers feel themselves under constant siege. Black cab drivers, eager to assert their ownership of London, cut them up or deliberately give them wrong directions when they get lost. But it's the passengers who are the real problem. Unlike the black cabs, and unlike the yellow cabs in New York, there is no partition to shield drivers from drunken shit-stirrers bloviating about the IRA or asking them: "So, have you met Osama then?" At the beck and call of up to twenty passengers a night, each of whom is heading to very disparate parts of the city, the drivers don't have a safety zone or a patch they can call their own. As the night draws on, and pubs start chucking out their clienteles, the cabs turn into wild wagons full of dishevelled, out-of-it Londoners who can't decide where they're going and don't want to pay when they finally alight.

Passengers may be deranged. Others just act as if they are. Mini-cabs, often used by drug dealers dropping off stashes to their clients in bijou flats and totterdown estates alike, are also popular with prostitutes, who may themselves be crack addicts. The women, who put on Eliza Doolittle accents when they step inside, soon revert to effing and gobbing when they're asked to stump up the fares. They cry rape. Better that though than the women who ask the cab drivers to stop off in dark back streets, run away without paying, then get their girlfriends to gang-rush and rob drivers who chase them to get their fare.

Areas of London that raise the blood pressure of cabbies include Dalston, Brixton, Leyton, Barking, Harlesden and Tottenham. African drivers rain curses on young Anglo-Caribbean kids, while Afghani and Pakistani drivers hate Asian customers on account of their refusal to offer even the most derisory tip. Many carry bats for self-protection. And many, even those who have been battered themselves, whisper with still-raw sorrow about the time their older, turbaned colleagues stumbled into the taxi office, whimpering and bloodied after being beaten up. "What sounds do I associate with the night?" ponders one driver. "It must be that sudden, tiny noise you hear when a knife is being unsheathed. And then I know that in a second's time I'm going to be feeling a blade pressed against the back of my neck."

"What do I like best about London? Nothing. Nothing at all."

In times past, immigrants to London worked in factories, café base-ments, small garment pressers. They led stationary, enclosed lives. Too tired or too poor to sight-see the city, they found themselves insu-lated from the new world to which they had travelled. For mini-cab drivers however, especially those who have just arrived in the capital and start out as anti-Knowledge blank canvases, every night is a crash course in unfolding urbanism. Every night is an education in extremity and juxtaposition. They soon become metro-savvy, worldly-wise and weary. They see pimp armies; packs of teenagers getting ready to jump an iPod-wearing schoolboy on his way back from a carol service; electronic shops being ram-raided; mass rucks outside boozy pool halls; tuxedo-wearing accountants staggering out of posh hotels by Hyde Park blootered after their companies' annual ball.

The night takes on the hard, flat quality of a video nasty. And the drivers become portable and manned CCTV cameras, logging and data-capturing the nocturnal city. They spy back-seat adulterous kisses, hear errant lovers reeling off fake itineraries on mobile phones to their partners at home, try not to stare at the stripy-shirted financiers fumbling with the bra-straps of the hookers they picked up in Clapton. They despise these girls who they transport regularly to 3am hotel-calls in Sussex Gardens and Bayswater; they lust after them too. Almost all of them have been offered sex in return for waived fares; not all have refused.

As the night rolls on, passengers, their tongues increasingly loosened by alcohol or chemicals, or the simple desire to unspool their biographies to someone whom they will never meet again, treat the drivers as confessor figures. Journeys at night tend to be longer than those in the daytime, with customers happy enough just to be on their way home to worry too much about the time the ride is taking. They're in comedown mode, reeling off sad and funny and self-incriminating stories. The drivers, though they affect to have heard everything before, get sucked in by these tales. They move, for a small while, from being mere people-ferriers to fellow travellers. They become passengers, prurient and complicit.

"You hear stories that are so sad. There was a gentleman, very smart, very respectable, who used to call in and ask me to take him home to Ilford. From a West London casino! £35! In the

back I heard him very often moaning and groaning: 'Oh dear god. Why do I do it? Why can't I stop?' It turned out he used to run a big department store in Ilford, and that he used to be a rich man. Because of the gambling he lost his wife and his house. But he was so smart; I was very curious about him. He always asked to be dropped off at the corner of a very dark street in Ilford. I was also worried he would be attacked. So one night, after he paid £35, I secretly followed him. It turned out he was staying in a Salvation Army hostel. He lost every month, and had almost nothing, but he could not stop.

Then there was the gentlemen who wanted me to help him to do away with his parents..."

Towards the end of the night, when they calculate they've made as much money as they think they're likely to in that shift, the cabbies leave their office for the last time. Their colleagues – London's basement brigade, its furtive enablers – will be sprawled on battered sofas, snoring perhaps, or silently watching satellite-news footage of Israeli army divisions bulldozing Palestinian settlements. It doesn't seem the right moment to regale them with the story about the obese customer who earlier got stuck while trying to get into the cab or the couple who became shirty when their offer of a fifty-euro note for a three-pound fare was declined.

The cabbies start their engines and begin to head home. They're tired and can barely stop themselves from nodding off. Tired of the murderous hours they keep just to earn the kind of money in one week that some of the passengers blow in one night. Tired of having to deal with lonely Londoners who call for a car simply to find a neutral space and a neutral ear into which they can offload all their anxieties and isolation. Tired of the mini-cab becoming a pressure chamber, an emotional landfill. It's almost too much for the drivers. Their cheeks stubbly, the atmosphere stale with many different varieties of nocturnal breath, they open their windows to let the air in – and the misery escape.

VIII

Fugitive Texts:

Graffiti Writers

FUGITIVE TEXTS:

GRAFFITI WRITERS

Antoine de Saint-Exupéry wrote of "Night, the beloved. Night, when words fade and things come alive". But he was wrong. Night time buzzes with chatter. It is saturated with words and with text. Fanzine editors and club-night promoters hover outside poky taverns handing out flyers to stripy-stockinged goth girls and their boy-friends. Puffa-jacketed geezers stuff telephone booths with garishly Photoshopped cards advertising the breast sizes and educational prowess of local prostitutes. Flyposting teams armed with buckets of potato paste ride around town in white vans, hopping in and out to stick up A3 sheets about forthcoming tours by young comedians. The radio waves crackle and fizz with the sound of a thousand venomous MCs spitting and freestyling rhymes from inside high-rise council estates.

And then there are graffiti artists. They scuttle through the streets like urban foxes, textual happy slappers ready to coat the city with elaborate spirograms of colour and slanguage, a floetry of illegal form and content. They are, they like to think, artful dodgers, lexical prestidigitators who operate under cover of night. Space invaders, action painters, smash-and-daub pirates, invisible and tersely monikered showmen who have chosen to use the city as a canvas for their rudeboy art. They thrive on danger, regarding the hazards they face and the risks they take before they get to whip out and start spraying their cans of aerosol as a turn on, a necessary precondition of their fugitive scripting.

All of them are addicts. Painting – not just the product, but the adrenalin-rush they get from the process – becomes their obsession. It takes a long time to build up a portfolio and develop a reputation in the capital; they need to plug away with the same metronomic dedication as milkmen or postmen. Only a minority have steady jobs; the nocturnal hours needed to get their name up in paints means they fall into bed at almost exactly the same time that most Londoners are catching the tube or bus to work. The few who manage to bag part-time positions do so only because it helps to pay for more paints and sprays.

Graffers may like to think of themselves as shadowy lone rangers, cloak-and-dagger textualists, but their social composition, as they will readily admit, is far from heterogeneous. Almost all of them are young men who started out as disaffected teenagers pumped up full of testosterone and a need to vent their inner turbulence, hungry for the kicks or kudos they could get by playing a sophisticated version of Knock Down Ginger.

These aesthetic cherryknockers are mainly white, perhaps not least because the police tend to crack down more heavily on black teenagers out on the town at night. Many are middle-class misfits – bishops' sons, privately-educated graduates, Lloyds' microfiche operators – social autists who find in darkness a confidence and eloquence they lack in daily life.

"The greatest buzz is when you've stayed up all night painting a carriage in a station that you've been staking out for ages. The security guards saw you and chased you but you managed to get away from them and hide. So you're there, hiding in a corner, totally knackered but you still really want to see your train. Then when it finally pulls in and it's got the design that you've been thinking and sweating about for weeks – that's lush."

The *London A-Z* means little to most graffers. They create their own maps of London. They see themselves as metro-rodents, part of a slippery squad of restless voles and tunnel rats who operate according to geographies darker and craftier than those of other city dwellers. They hang around bridges, empty playgrounds, arches full of smackhead debris. They are drawn to the derelict and dingy: broken yards, battered lock-up garages, semi-deserted industrial estates – spaces where they hope to find the time and the peace to paint unhindered by the law.

Abandoned, obsolete London is their fiefdom. No part of the city is antiquarian to them. Over time they build up a vast, working knowledge of its disused train lines, secret tunnels, the catacombs below Farringdon, the shafts at Brixton. They know how to break into buildings, which doors are alarmed and which disabled, how high a drop from a window will be. They ascend higher and descend lower than other Londoners, climbing under fences, shinning up pipes and stomping across rooftops with an ease that seems to ridicule the night's murky light. When they get together, as they occasionally do, they swap stories about pals who were chased by rifle-wielding vigilantes, electrocuted on live rails or had their arms ripped off by oncoming trains. They'll talk about 150-year old sewers that they managed to penetrate, the postal tunnels that MI5 operate from Mount Pleasant.

Graffiti artists are speed-orienteers and routefinders. They compare themselves to dyslexics who can mentally rotate the capital in 3D. "We see the city as a shape," says one, "We're always looking at it from above rather than through the dimensions in which we're walking." Theirs is a mutable cityscape, its architecture a set of conundra and challenges to be overcome. They divine ways in and out of buildings. They can gauge the weight of a drainpipe down which to slide. They know that if they cut through an alley and through someone's garden they will end up in the same place they started.

This mastery is not total:

"Once we were in West Hampstead trying to climb over the back of this notorious kebab shop that used to poison everyone who ever ate there. It was winter and the place was covered in frost and ice. My mate, who was quite big, walked over some wooden planks near some undergrowth. They gave way, so he fell into a septic tank full of meat cuttings and dripping oil and piss and sick. It was freezing cold too but he still couldn't scream, otherwise we'd get caught. The last time we saw him was running down the road on his own, bits of kebab falling off him."

Graffers are criminals. They wouldn't have it any other way. To legalise their work would be to delibidinalise it. They are willing to stake out a prime site for as long as two months, usually during the winter when they have more darkness to play with, giving up their weekends so that they can lurk like private cops logging the movements and tabulating the itineraries of security staff. They are stalkers, burglars casing a joint, combatants bent on storming military patrol. The language they use for their handiwork – 'bombing' involves writing your name on as many places as possible; a 'throw up' is when a long name is abbreviated – conveys the damage and messiness that they wreak.

Uniforms, or any costume worn for working rather than partying, help them stay undercover or invisible. Those who wear the orange coats of railway workers or the bright yellow waistcoats of street cleaners find that no one gives them a second glance. Others don helmets, balaclavas or doo-rags. A melodramatic and health-conscious few wear gas masks to stop them inhaling spray-can fumes that scald their lungs. They'll also come armed with boltcutters, washing-up gloves to protect their hands from give-away traces of paint, and polyester overalls that they ditch easily if caught.

Veteran graffers and taggers don't expect to get caught too often though. Their ears are alert to the sound of police-radio crackle or the distant crunch of Dr Marten'd security officers. And because they are, in however soft a fashion, part of the night's criminal fraternity, they don't feel like victims. They develop neutral walking styles exactly half-way between over-confident flexing and the shaky vulnerability of the drunkard or the homeless. They don't even care about roaring helicopters or the thousands of CCTV cameras festooned across the city. The former, they believe, regard them as small fry; the latter are very often broken or unmonitored – mere ventriloquist surveillance. Karaoke panoptics.

"There's a mystery to graffiti that is really appealing. It can challenge your visual sense. You ask yourself how did someone manage to get up that building to write that? There's no drain-pipe and no ladders that tall. It's weird. I remember seeing some along the Thames, on the South Bank. I thought it must have been written by someone standing on a boat, but then I found out it was actually someone hanging on a rope off the side of the bank. Even though it was a low tide, he was getting his feet wet all the time he was painting."

The micro-texts that the graf artists create are pieces of ephemeral street furniture that fill the areas where they are found with fragmented reminders of a forgotten history of post-war pop writing: from the KBW signs scrawled by anti-immigrant bully boys and the chads chalked by medical students in the 1950s, through the anti-Vietnam and George Davis Is Innocent slogans of the 1960s and 1970s to the pro-Ocalan or Bin Laden-exalting messages daubed by second-generation ideologues today.

The tags and throw ups, by their very existence as much as by their words or their obsessively rehearsed shapes, are less a dirty protest against modern London, and more a fierce yelp of freedom and straight-to-hell kidult affirmation. They are part of the chatter, the spectrum interference, that city authorities feel obliged to silence. They tattoo the skin of the city; disfiguring or, according to the perspective of the viewer, beautifying it. They say what so many people who work at night would like to shout from the top of Telecom Tower: "I am here."

Graffiti is a kind of fingerprint. Metropolitan police officers have been known to turn up to the opening nights of graffiti-based gallery shows armed with photographs of tags in order to pin down the individuals behind the art. Veteran taggers don't need to resort to entrapment; they claim they can read off a biography from most walls. The height of a text betrays how tall the artist is; the slope of the lettering if they're left- or right-handed; the intensity of the spray whether the can has been tilted or held at length.

Ornate styles, those where the tops of letters have a lot of flare, or which effulge with star and cloud symbols, tend to be the work of Europeans. British tags, by contrast, are normally more blunt, less affected: the can is neither pulled back or pushed forward – merely held straight to the wall. London tags are noted for their rawness: almost always simple motifs and on the small side, they are often in black and white as those putting them up don't have enough time for colour. In villages or small towns, it only takes one or two distinctive tags to make an artist's name; in the capital, where competition is fiercer, there is a greater focus on quantity. The art veers towards branding: the shape of the letters, endlessly repeated across the city's buildings, is as important as the words.

Graffers, relentlessly combing the city looking for next surfaces and sites to paint, are keen students of typeface and texture. In their spare time they flick through specialist books and jackdaw the web for images drawn from Mexican wrestling masks, 1950s pulp cartoons, fetish zines and sixteenth-century Bavarian tombstones which they can recycle or recraft for their own designs. At night, when they're not blasting the walls with paint-filled fire-extinguishers, they clock the signs and calligraphies of the city with connoisseurial eyes. They notice, even in dim light, that handpainted storefronts are disappearing, only to be replaced by corporate chains that sport logos familiar from television and newspaper ads. They notice that space previously occupied by fly posters and student xeroxes is being bought up by Clear Channel.

Graffers also know, first hand, that London is starting to feel different. Alien to the touch. Electricity boxes used to be green and smooth; now they're stony and spiky to stop them being defaced with marker-pen insignia or stickers. Anyone who brushes against them ends up as grazed and scratched as if they'd just emerged from a playground fight. Other public surfaces have tough vinyl coatings, with the steel designed so that any illegal paint can be easily blasted off.

Some taggers see themselves as playing a cat-and-mouse game with the law and are excited by the challenge of having to hatch new ruses to counteract the emulsifying strategies of local councils. Others view such crackdowns as creative straitjacketing, burgher-pleasing three-line whips designed to snuff out anything messy, random or spontaneous within the city. "All the local authorities have anti-graf squads now," complains one. "Plus they're spending quids on fat forty-year-olds in Japanese vans with high-pressure hoses to rid the streets of chewing gum. Everything is getting cleaned up and buffed out. It's just a different version of them tagging all these young kids and sticking ASBOs on them."

Graffers, whether they like it or not, are part of the urban grime economy, enemies of order. Their art is deemed a fist of defiance. They themselves see, night after night, how the forces of beige-ness are encroaching on the spaces that they used to encroach upon. The warehouses and abandoned shops they lacquered with colour are being razed and reconstructed anew. Empty, weeded sites are being filled in. Dark streets are better lit since well-connected newcomers to formerly tough areas hassle their councillors about improving local safety. Graffers, struggling to lurk in the city's disappearing shadows, know that their art is migrating to glassy surfaces and onto the web. They also write in the knowledge that the Olympics will soon obliterate their texts: the Games' organisers, keen to ensure that tourists clap eyes on a gleaming, yacht-white capital, have pushed through zero-tolerance policies that means graffiti disappears within 24 hours of going up.

"Once we were on this rooftop near Hoxton doing a painting job. Then all of a sudden this woman who lived over some shops stuck her head out of a window right by us and started shouting: 'I know what you're doing.' We thought she was going to tell the police and were ducking down. Then she said, 'Don't worry. It's cool. But the tiling's not safe. When you finish just come and knock on my window and I'll let you out of the door.' We were a bit surprised..."

Graffers create liquid architecture, temporary graphic structures that, no matter how much graft or guile has gone into their production, may vanish within hours. Their self-proclaimed works of vandalism may in turn be vandalised by other street-artists playing at one-upmanship. They try to insulate themselves from regret by taking snaps of their work, leaving home in the evening with fresh reels of film in case they get caught by the police who use old pictures of them larking about with their crews to track down and prosecute their friends. A painting is never really complete until it has been shot and entered into the artist's noctographic portfolio.

Some parts of London, however, are cleaned less rather than more often. Artists talk in hushed, reverential tones about a tunnel from Moorgate to Farringdon that contains a huge archive of metropolitan night texts from the last twenty years. Sunlight, even more rain, destroys wall markings; darkness freeze-frames and preserves them perfectly. Wipe away the dust, point a torch at the tunnel wall, and there, for anyone hardy enough to smuggle themselves into this subterranean gallery, are markings and tags written in styles dating back to the dawn of London's urban graffiti scene. They are as thrilling for the graf artist to unearth and behold as prehistoric cave drawings might be to a historian. In the ephemeral world of night writing, these B-Boy calligraphies are almost Palaeolithic.

Walls have appetites. They soak up text, draining it of colour and definition. But this process can take many years. Artists talk of ghost walls on which the outlines of old shapes and designs can just about be discerned. These are rebukes to the aggressive, now-fixated amnesia of some night writers. They also offer the opportunity of applying one layer of text onto another, of creating a tacit dialogue between past and present, a graffiti palimpsest.

"Sometimes you get a shock when you see homeless people in tunnels. But they're scared of us as much as we are of them. Mostly it's workers you see at night. If they don't turn their dogs on you, they'll give you a cup of tea and talk about the old days: 'When I was your age we'd be throwing stones through windows'. Then when you tell them you have to go, they'll say, 'By the way, mind you stay off the rail,' or 'Carry on a hundred yards after that corner, and there's a big bridge you'd like.'"

Graffers are so used to being labelled vandals that they learn to embrace the tag. They revel in the fact that those outside their circle think of their painting as vomit, artistic flytipping. But they're far from blind to the beauty of what they do or of the city on which it is displayed. Summer, though it allows them less hours to paint, is when they go to work in shirts and shorts. They'll carry 4-metre long rollers that they lug up to the roofs of tower blocks and then lean over the side and write their aliases upside down. The fumes stink and soon they feel the twinge of tendonitis.

Finally, just as dawn is breaking through, the job is completed. Their feet still dangling over the edge, they'll reach into their rucksacks for a can of beer or light a joint. They feel a woozy tranquillity and a pride at accomplishing their latest mission. The sun begins to come over London and gently lights up St Paul's, the Gherkin, the Nat West Tower. Below them are the infinite streets and back walls which, as tongue-tied, acne'd teenagers many years before, they started painting to make the girls at school like them. Now, just for a second or two, the city feels all theirs. They smile and nod at this thought. "Nice," they say.

IX

Ale Always Tastes Good:

A Thames Barger

ALE ALWAYS TASTES GOOD:

A THAMES BARGER

Out on the south shore of the River Thames, east of London Bridge, beyond the Barrier, farther even than the garbage islands of Rainham Marshes, lies Northfleet Works. Opposite is the Port of Tilbury where, in June 1948, the *SS Empire Windrush* berthed to allow almost 500 Jamaicans, part of the first wave of post-war migrants from the Commonwealth, to head towards central London. The Works itself has no such claim to distinction. Scheduled to close soon by its owner, the cement manufacturer Lafarge, it already appears to be sinking into the bogs and dour wetlands on which it sits and which will soon be developed as part of the multi-billion pound Thames Gateway regeneration scheme.

The site, known to those who still work here as The Dusthole, is semi-derelict. The timber has rotted, the machinery is rusty and creaking with old age, the elevated rigging sheds are broken. Weeds threaten to overrun the quay. The air is thick with the waft of hot dust. From here, almost every evening, in summer and in winter, a 240-ton Dutch motor barge sets off to carry cement up the river to plants in Battersea, Vauxhall and Fulham. Twenty years ago, there were a hundred motor barges going up and down the Thames; now there are eight.

The skipper of one of these barges, *Gabriele*, is a beaming 62-year-old man called Alan Jenner. He has been working the river for half a century. Towelling off the dust and sweat from his muscly, ginger-haired arms, he recalls that when he was a kid there were so many boats on the Thames that it was possible to skip from boat to boat all the way from one side of the river to the other without getting wet. His first job, as a twelve year old, was to row "ladies of pleasure" to the sex-starved sailors aboard steamers in the port of Rochester. They would throw him ten-bob notes and half crowns in gratitude.

"I'd like to be buried in the Thames. Some of my friends have done that. With one particular friend, we threw his ashes over the side of the boat but they came back straight away. The wind blew him back. We'd thrown him the wrong side. Someone said: 'He's coming back for a bit of overtime.'"

The men who work on the river are a dying tribe. They inherited their jobs from their fathers and grandfathers. The docks, for all that they seemed exotic and polyglot to landlubbers, was home to clannish and tight communities that in recent decades have been unmoored as the capital, or at least those in charge of it, decided to abandon its industrial heritage and transform itself into a tourist citadel and global financial hub. The river has become the preserve of pleasure-seekers in buzz boats, cocky young gastro-revellers going wild on corporate credit cards, ruthlessly orchestrated photo-opportunity shoots for imported Ukrainian soccer players, enthusiastic out-of-towners rushing to the prow of *HMS Belfast* to mimic Leonardo DiCaprio's 'I'm the king of the world' speech from *Titanic*.

Those that are left from an earlier time see themselves as survivors. Wilting survivors. They move through a river that appears to them to have been razed and colonised by outside forces, its soul abducted. They look around, gazing, mystified and sometimes trembling with bitterness, at a ghost architecture of decommissioned power stations, wharves that have been torn down in order to make way for storage spaces, warehouses that have been converted into apartments for bankers and designers. The tankers and cruise ships that they pass sport the insignia of foreign companies. There is only phantom industry now – the cast-iron buttresses of Beckton Gas Works that are too costly to be dismantled, a few faded inscriptions on the sides of flour mills that have been kept to add value and the patina of history. The shore, which used to have public landing piers for anyone to dock and disembark, has been sold off to private companies.

The bargers laugh that the new riverside dwellers, when they're actually resident in their luxury pads, and not swanning around on foreign beaches or ski resorts, are always agitated. Many of them, regarding the Thames as mere wallpaper, a toney backdrop to their manicured lives, had bought their flats unseen and are now always on the phone to their agents complaining about the noise of boats going past or because they can spot power stations from their balconies.

In the eyes of the bargers, it is the flats that are the most visually disruptive. They describe them as "luxury prisons", domino-sized cells within bulbous high-rises or brutally angular developments that are garished up with strips of blue and pink neon, bolts of Vegas-chintz lightning. The pavilions, faux-pagodas and Riviera-style hotels that huddle together near Chelsea they describe as "high-class knocking shops." What, they wonder, does London have to offer to nurses, policemen and dockers: "How good can free enterprise be if it's causing poverty for so many people?"

"Me and my missus had to divorce on the grounds of political economy. My wife had osteoarthritis of the spine, and then a lesser version of ME. Even though she couldn't work, St Thomas's wouldn't put her on their priority list for treatment. So we had to get divorced. I mean – she kept falling over and was always full of bruises. I have to say, if my barge had as many defects as she had, I'd scrap it."

There is a Dutch sign that hangs in the wheelhouse: "Al is er storm/ Of tegenspoed/ Een borrel smaakt/ Altijd goed". It translates as: "Even though there are storms/ Or bad times/ Ale/ Always tastes good". But neither Jenner, nor his younger relief-captain Dave, nor his teenage deckhand Luke, drink on board. They prefer more homely fare – crisps and chocolates and PG Tips – carrier bags of which Dave, who has just returned from a nearby Sainsbury's, winches down on a pulley from the wharf to the boat. Before they can start tucking in, they have to ensure that the pipes of cement are properly positioned in the cylindrical tanks. This can take a couple of hours: the barge's load of 240 tonnes is a dozen times greater than that of the lorries that transport the material by day.

By 9 o'clock *Gabriele* is ready to leave. Dave has gone below deck for a quick snooze; bargers run a tight operation, and have to subsist on three hours of sleep a night, snatched pockets of shut-eye. The sky is inkwell-black and the seagulls are mute. There's no slish of traffic from nearby roads. No patrol boats ping by. Jenner rubs his hands with glee: he dislikes the tourist boats that clog the river by day and zip around at dangerous speeds. Even though a night like this is restrictive, its lack of light making navigation harder and the use of radios, depth gauges and radar systems crucial, he prefers its quiet company to that of daytime.

Jenner trusts his eyes more than he does any technical equipment. For over forty years he has been building up a film archive of the Thames, snapping and shooting the creatures, boats, buildings and foliage that he spots as he chugs up and downstream. He believes there is an art of looking at the river, one that can only be developed slowly over time, and that is impossible for those youngsters who dwell on the job's derisory pay to fully appreciate. The way he sees it, each evening is a completely different journey: breezes will make waves behave differently, strange aromas will drift across the river; the play of shadow and light will throw up bizarre shapes. "They're like a half pack of dinosaurs rolling around in a heap of sand," he says as he peers out at a huddle of cranes on the opposite bank. When he was younger he had a girl-friend to whom he confided his passion for taking photographs of dockside machinery: "She said to me, 'Are you some kind of nut-ter?' Well, there was no point in having a relationship with her."

"If the souls of dead people came out of the Thames now they'd dive back down again because they wouldn't understand what they were seeing. They'd say: 'Where's our mud hut?'"

Bargers are weathered Londoners. They don't have double glazing or air conditioning systems to insulate them from inclement condi-tions. Fog, although it doesn't last for three days as it used to back in the peasouper era, can hang around for eighteen hours and prevent them from setting off. Mostly it arrives without warning, a grey blanket flung over their faces that forces them to stop once they're already on the river. Storms, however fierce, are easier to deal with. As the bargers turn and twist through bends in the river, its increasingly surly waters dark and colourless, they can see vicious squalls raging miles ahead of them, or heavy snow lighting up the radar screen. Then, one more bend, and suddenly they find themselves in the middle of a gale. A big one can kill winds, and leave the river quiet for up to a day.

Oversized mosquitos fill the air in the earlier part of the evenings. That's what the bargers call the private helicopters roar-ing above their boats on their way to the rooftop helipads that are increasingly visible from the river. They also see and feel jet engines flying through the clouds, causing, they are convinced, huge and potentially disastrous pressure changes to the atmosphere. Jenner, who goes to air shows in his spare time, swears that he can tell which plane is flying above his barge just by the way that the tea shakes on his table.

In the summertime, real mosquitos settle on and munch their flesh. Midges swarm and bite. But most of the animals they see – and even those that they don't: jokey speculations and reveries about the whale that got stuck in the Thames in January 2006 kept them warm throughout much of the following winter – they find delightful. Seals. Dolphins. Rats they claim are the size of Jack Russells. Porpoises at Fulham that swim right alongside the barges, seemingly in thrall to the squeak of the cutlass bearing in the revolving propeller shaft that may or may not sound to them like mating calls, and keep disappearing under water before re-emerging all the way downriver at Blackwall Tunnel.

"Nobody knows we're here. Nobody. The Port of London sent a river pilot to assess my competence. I've been on the Thames for fifty years. I was the one who trained that pilot."

Working nights is tough for those bargers with girlfriends or wives ("Have you emptied your balls this weekend?" Jenner likes to ask his deckhand on Mondays, knowing that he won't be getting any more action until the next weekend). But at least they don't have to hang around at wharves waiting for lorries to be loaded up with cement. The river, except on the choppiest evenings, is peaceful. The bargers, especially after midnight, feel as though they have been unshackled from the city, its soot and heaviness, its noise and overbearing solidity. They breathe in the fumes of freedom, bathe in the tranquillity of the dark waters through which they gently move.

And yet, perhaps because this atmosphere is so appealing to them, they cultivate a bleak and often racially inflected repertoire of stories about what crimes and knavery befall those who have the misfortune to live on solid ground. Rotherhithe: "You expect trouble down there. It's like watching *Big Brother* on television: a mad, messed up reality show going on before your very eyes." Southwark: "There's an estate there. The kids throw all sorts at us –

plastic bottles, biscuit tins, bricks, pennies." Wandsworth: "They're cannibals there. You hear these screams and groans. Blood-curdling noises. Do I dare risk wandering around Wandsworth at 5.50? If I did I'd be reading my obituary in the paper in the morning." Brixton: "As far as I'm concerned, they can just put a bubble over it and fill it with gas."

These are not just tall tales. Some of the bridges along the Thames may be cammed up, but that doesn't deter teenagers – and also their pissed fathers – from spitting at WI members enjoying the riverscape. It doesn't stop armed robbers breaking into bankside alehouses. Jenner's barge leaves from below Gravesend, the mournful starting point of Conrad's *Heart of Darkness*; sailing upriver, the shoreline dimmed and obnubilated, it's easy to conceive of London as a barbarous jungle-continent full of mindless hordes drinking and chanting and dancing to electronic-tom-toms, and flanked by foliage-camouflaged garrison towns to repel naval invaders.

"A woman went to get a taxi with her son at the Isle of Dogs. At the street corner it was all whores hanging around waiting for trade. 'Mum, mum: what are all those women doing?' Mum was embarrassed, but quick-thinking: 'I expect it's the sailor's wives waiting for their husbands to come back from their ships.' The taxi driver leaned back: 'Don't give him any of that shit, love. They're whores, that's what they are. Whores.' The boy says, 'Mum, do whores have babies like normal women do?' 'Of course they do, Tommy. Where do you think taxi drivers come from?'"

Londoners take the Thames for granted. They may walk across its bridges, but they rarely sail across the water itself. To them, just as to the location scouts who make sure that every Hollywood movie set in the capital has at least one shot in which A-list star-crossed lovers wander its banks with a view of St Paul's Cathedral behind them, it's seen as a place of lazy fun. The bargers, though they love the river dearly, are closer in opinion to H. V. Morton who believed that the Thames at night was the most mysterious thing in the whole of the capital: "So much part of London, yet so remote from London, so cold, so indifferent...."

For the bargers know that they if they listen carefully they will hear the cries of thousands of stricken Londoners sinking into the turbulent waters: Elizabethan lightermen whose boats fell apart; eighteenth-century African slaves jumping overboard to avoid being deported back to plantation servitude; the six hundred passengers who drowned in 1878 when the paddle steamer *Princess Alice* collided with another ship at Galleons Reach, the victims of the *Marchioness* disaster in 1989.

The river is a place of death and disappearance. Bargers have been known to curse eco-vandals who fling bags of rubbish into the water, only to realise that those bags were human bodies. They have seen drunken revellers shout 'Wa-hey!' and leap into low water with the result that their feet got trapped in the mud and they drowned still standing. They tell of down and outs, misery-sodden tramps and OD-ing kids jumping into the river, unspotted by bystanders, so that their weed-entangled bodies are only found many days later.

Rumour spreads fast by water. The talk of the river is not the discovery of drug cargoes or of the headless torsos of sacrificed African girls. These days, it is the importation of illegal immigrants that keeps the bargers gossiping. The modern descendants of the welt-backed slaves brought to London after the sixteenth century are Eastern Europeans and Middle Easterners who are squeezed by traffickers into near-airless tin-box containers welded to the bottom of foreign ships. The containers, and the corpses inside, are only discovered when the ships harbour at dry dock. "Poor bastards," sigh the bargers.

"It's part of me. It's running through my veins. I dream, live and work for the Thames. I hope it is a life sentence, I really do."

The night floats by. Its soundtrack is the gentle chop and chunder of water, the metallic gurgle of the barge's motor, and periodic radio bulletins to the officials at the Vessel Trafficking System based in Woolwich. It wasn't that long ago that skippers had to make do without radio, or radar, or power steering; they just shouted across to the other boats. But then, it wasn't that long ago that the nocturnal river was swathed in blackness; now, even at its farthest reaches, carparks and grand shopping complexes are sprouting up, their light leaking out onto the Thames and denting its darkness.

As we move towards Greenwich, and then to St Paul's, new apartments dazzle with gay-liberation and graffiti-bright colours. The city's skyline has changed, the church spires and cathedral domes that gave it spiritual elevation supplanted by blobs and beehives and trout-pouted constructions seemingly imported from Legoland. The bridges gleam like candelabras. The Dome, built on poisonous junkspace from toxic strata of acid, coal and asbestos, still looks paralysed and absurd, an upturned crab unable to move. Canary Wharf, stiff and bemoneyed, its uptight verticality in contrast to the river's shifting, curvy horizontality, blazes out light: a bonfire of London's soul.

On we go, past the old holding cells near Whitechapel where convicts waited for the ships that would deport them to Australia; past the Marine Support Unit patrol boats whizzing by to find some unfortunate who has stumbled into the water; past the Embankment along which cheering partygoers have created a huge Conga line; all the way to Fulham where, after the boat's contents have been transferred to a cement plant, there is just time for a quick kip before the barger sets off upstream back to Northfleet.

We are tired and quiet, but as dawn slowly leaks into view, it is impossible not to be beguiled by the river. There are no traffic jams, no road works closing off access: just a long, sun-glazed vista that stretches for miles. The distance between us and the concrete city is greater than if we were travelling on a motorway. Just for an instant, but an acute and painful one, the buildings on shore look like holding pens, shackling devices to destrict the aspirations of their inhabitants who are mere lab rats running round in ever decreasing circles. The water is so calm that it seems as if it is the city that is floating and that we are the solid ones, the custodians of the capital's history, confidants to its majestic melancholy.

Here, on the Thames, for all the sweat and hardship, for all the worries that they are becoming posthumous in their own lifetime, the bargers can exist beyond the stiff rhythms and stressful schedules that govern the lives of most workers in London. Here they operate according to tidal time rather than to clock time. It feels intoxicating to them. They laugh a lot as the morning creeps forward: at the erratic movements of a party-goer returning home late; at the wobbling midriff of a City jogger crossing one of the bridges; at the crowds of suits milling towards town. "I see them and think: if one falls, they'll all fall. They're like little robots. You have to laugh." But I don't laugh. In fact, I feel like crying. Whether it's because of the river's brittle, pale beauty at this time of morning, or because those robots remind me of myself, I can't quite decide.

X

Contagion Set Free:

The Urban Fox-Hunter

CONTAGION SET FREE:

THE URBAN FOX-HUNTER

The evening starts with a bleep. In the dark of the cinema auditorium I fidget around for my mobile phone. There is a new text message: "The shoot is on tonight. Meet at corner of Hayes Road and Southall Lane." Immediately I scramble for the aisles, eager to get out onto the street and make my way to West London.

I am hurrying because I'm due to attend my first urban fox shoot. The black-cab driver says he wishes he could join me; he'd heard on the news that there were meant to be more than 10,000 foxes prowling the city's streets. He himself spotted a pack of fifteen of them while waiting for a client at Smithfield Market at 2 am the previous week.

This night has taken many months to organise. Bruce, my guide, had said he would be happy to talk about his job as the capital's premier fox culler, tracking and taking out thousands of them each year. There had even been a photograph of him in one of the tabloid newspapers, togged up in a camouflage jacket and armed with a power-rifle, squatting proudly over the corpses of 23 foxes he had shot in a single night. But the photo had got him into trouble: pest control experts argued that it was impossible that so many foxes would congregate in the same area; reporters pointed to Bruce's criminal past, digging out details of the six-month suspended sentence he had received in 1988 for shooting a teenage boy with a 50,000-volt stun gun.

Bruce had lain low a little after that, occasionally scheduling get-togethers, only to cancel them at the last minute because of a spot of rain. By the time the cabbie drops me off near a desolate ring road it's 10pm and I'm pretty sure he won't be showing up tonight either. An all-night Sainsbury's gushes out orange neon through which SUV-driving customers move on their way out of the suburb-sized car-park. Bug-eyed stoners, lean as syringes, jerking as if they've been electrocuted, lurch across a gas-station forecourt, nabbing customers and reeling off snot-caked stories about how they never got on with their parents and their girlfriends ditched them and a mate's phone is on the blink and could you lend us a fiver? Behind the plexiglass, an Indian till-clerk looks on with disgust.

"People assume that killing foxes is cruel, but the way I look at it is this: it's like going to a restaurant. You're there in the dining room having a starter, looking forward to the main course, sitting there with a really nice glass of wine. The next second? All over! You don't feel any pain or any danger or see anything that's going to happen to you. It's not like you have to crawl into a ditch to die. Same with the fox: he comes out to eat, and the next second: he's stone-dead."

When I find Bruce, he's standing by his Range Rover in the carpark of a sprawling sports centre. The golf course that surrounds it has been invaded by foxes. Amateur players, retrieving miscued tee-shots from the rough, have leaped back, startled to find a pair of dark vulpine eyes peering at them. The greens have been scuffed and damaged too. Bruce and his two marksmen assistants have been called in to exterminate the foxes under cover of darkness. They wear camouflage gear and big boots. Bruce, who is bald and has the stocky physique of a journeyman wrestler, is busy stamping on some rabbits that he shot earlier that evening. Already he's strung them to the car's rear bumper, crushing their intestines with his heel so that their scent will attract foxes. It's an unusual night; most times, he baits the foxes with defrosted chicken drumsticks which he buys in packs of thirty from his local Iceland.

Bruce was brought up in the country. He started shooting when he was 16. He's 47 now and the log he keeps tells him that he shot over 1400 foxes last year. In a good week he'll trap seventy. Unlike other pest-control firms whose staff regard themselves as conscientious objectors and whose bosses fear that their vans will get attacked by protestors, Bruce's advertises itself on the basis of its long-established track record of culling foxes.

Each year the numbers are rising. The urban fox used to be an exotic creature, worth serious currency in the city-rumour stakes. He was a metropolitan Big Foot, a pavement Nessie. Now, as London expands, and the division between the city and the countryside becomes blurred, foxes are commonplace. They represent the return of repressed nature to a spayed and neutered capital. They battle down upon us because, whether we know it or not, we are wrecking their homes and whetting their appetites. They are the trace of the human.

And so they creep into the city, sharp-toothed soft treaders who are drawn to its cemeteries, industrial estates, overgrown gardens, public allotments. They are contagion made solid and set free. Black-economy thieves. Illegal aliens. Immigrants who feed off the capital's excess, its scraps and remnants, all the while inhabiting precarious, marginal lives out of view to most Londoners.

Urban foxes, like urban rats, are stealth intellectuals. Each year they get tougher and tougher. They become hardened to the blare and roar of the city. Impervious to car fumes and to chemical or electrical smells. One meal can sustain them for a week; if they have four or five in a short period they develop energy to burn, and will lark around doing damage to gardens and to private property. They also thrive on challenges. A bait or trap left for a rat might stay untouched for days while the rodents ponder and cogitate; foxes though will regard them as a challenge. This high-risk strategy means that five traps left on a school playing field may yield a full house of five foxes the next morning. Equally, if one is caught, a vixen will show her cubs how to avoid danger in the future: she'll physically pull the trap and go off and dig beneath it before sliding her paws in at the side to retrieve the meat.

Short of a rabies epidemic hitting London, the urban fox is here to stay.

"We've missed our vocation. We should have been snipers in the armed forces."
"Why didn't you?"
"There's nothing in the country worth fighting for. Unfortunately. The way everything's going..."

The foxes who move through London at night aren't all diseased and scrofulous. Some are more Raffles than Artful Dodger, gentlemen thieves with a taste for the high life rather than sooty street-knaves. These gourmand foxes gravitate to Hampstead, Fulham, Kensington and Chelsea, places where the gardens are manicured and expansive, and the leftovers are from the finest, most protein-filled cuts of meat. Their fur, when they're finally gunned down, will be sleeker, redder, more textured than that of their prole brethren rooting around in the dustbins of East End council estates.

Fox hunters love these areas too. The owners of the grand homes in Kensington Palace Gardens and the staff in the palatial West London embassies treat them with respect. Upon walking in, they are offered three-course meals served up on silver platters. After a glass of vintage wine (only one glass – the technicians worry that they might get tipsy and either fall asleep on the job or lose accuracy while shooting), they go upstairs where they are provided with a pair of bedroom slippers and a cushion for them to rest upon while they sit on the window sill waiting for the moment when they can blast away at the foxes. It certainly makes a change from having to squeeze into smelly vans for long stretches on wintry nights.

Areas that fox hunters don't like include Battersea, Croydon and Stratford. The councils here are less effective at organising regular culls. The foxes are more likely to be emaciated, rat-tailed and riddled with mange. They smuggle into the piping underneath the foundations of tower blocks or squeeze themselves into high-rise catflaps so that the flats get infested with thousands of fleas and ticks. East London foxes tend to be small; one vixen caught ripping through sacks at a Romford warehouse was a mere twelve pounds, while a fox found in plusher Sutton was, at 32-and-a-half-pounds, almost the size of a collie. They are said to be noisier too. Those in Brixton and Finsbury Park squeal and shriek so horrifically that, according to Bruce, "it sounds like a baby crying or a woman being strangled to death."

A hunter tells me of a shoot he carried out at a three-floor Victorian house in south London. In the garden to one side were Australians and South Africans enjoying a barbecue while a dog ran up and down; on the other side, a couple was having a meal outside while listening to classical music. Neither sets of individuals knew that at the end of the garden a few yards away from them, two foxes were eating set bait and were about to be shot in the head.

"We go to fancy houses and we go to shit tips," Bruce says. "We are like undertakers. There will always be a need for us. We have strong stomachs. We have to. When a fox dies under the Porta-kabin of some secondary school you have to grab it, but half of its body might rip in your hand. Then you have to go and scrape up the rest of it. The smell is pungent and sickly. You never forget it. It will get through your mask no problem."

"I did have a fox once: Butch. He was tame. He was our pet. He used to live in our dog run with our Jack Russell terriers. They

thought he was a Jack Russell terrier too. Funny, they'd go out to hunt foxes, and work on getting them out of buildings. Because they hated foxes with a venom. But then they'd come home and Butch would be in the dog run and greet them and be all submissive to them. Then, when he was twelve, he got out and when I came back home he'd been shot by the farmer across the road. I was gutted."

Fox hunters admire the ingenuity and hardiness of their prey, but they are never sentimental about them. They regard each vixen they see as a poisonous aggregator, the potential mother of six or seven vulpine terrorists. They reel off sob stories that justify their venom: one is of a Koi carp owner who had spent £15,000 building a 20-foot pool for his prize-winning fish, some of them worth £3,000 each, only to wake up one morning to find the shredded remnants of their heads and guts lying by the poolside; another is of a 96-year-old Welshman who came to Islington after the war to live with his wife whom he had meet when she had been evacuated to the Borders. When, many decades later, she died, he turned their garden into a shrine to her, a tranquil memorial where he could solace himself with memories of the years they had spent together. Then, one night, a family of foxes had wrecked its lawns and flowerbeds. He felt defiled.

Urban fox hunters are mercenaries, revenge artists. They act out Londoners' darkest fantasies. City dwellers habitually dream of being Travis Bickle, of finally saying no to the noise, terror and hysteria that threaten to engulf them, by going out, weaponry in hand, to slay the source of their misery. Fox hunters always deny that they are Barbour-jacketed robo-cops or chicken-drumstick-carrying terminators, and talk at great length about the precision training they undertake in order to carry out their commissions safely. None the less, at a period when Londoners often lament the inability of the police to capture yet alone punish robbers and street thieves, the symbolism of the hunters' jobs is unmistakable and, to many, very appealing: search and destroy.

Hunters especially love the occasions when they're called to help out those people they call 'bleeding heart liberals'. They mean those men and women who for years had shouted their hatred of fox-hunts and muttered sympathy for activists who slashed the tyres and broke the windows of marked pest-control vans. Bruce says he has very little time for the moccasined nobbs and toffs of the Countryside Alliance who took over the streets of London to defend rural hunting because they regard it as a way of life rather than a way of making a living; yet, though he tut tuts and shakes his head, he smiles inwardly when he is called to assess the damage done to the gardens of pro-fox sympathisers, landscaped to the tune of £40,000. The lawns will have been dug up, the irrigation and self-watering systems ripped, the safety lighting easily cleared.

At times like these, the hunter feels, however fleetingly, a bond with the hunted. He sees the fox as a fellow night-traveller. A worthy spar as much as an enemy. A jouster for mastery of the city's nightscapes. A dark tactician who lays up in woods by day, their messy foliage serving as temporary hostels, green rooms before the centre-stage entertainment that ensues after night fall. In response, the hunter turns the imperilled garden into a castle. He needs to: even a ten-foot wall with a buttress holding it up is insufficient defence; foxes are just as agile as cats, and need only to get a grip with one paw to soon be up and over. They can run at speed along fences with 1½-inch rails on top of them, and dig and burrow three feet below them.

"I'll never forget the worst fox I ever saw. In its back leg it had bad maggots the size of those you'd get in fish bait shop. They'd got right to the bone in his back leg. It was absolutely rotten. It

stunk. The poor thing was limping around. I shot it to save it from a lingering death. I've also shot foxes before now that I keep thinking: 'He's such an unlucky fox.' One that's had his leg broken badly, or was born with two legs shorter than the others, so that when you see it it looks like a sort of thalidomide fox..."

By the time Bruce and his colleagues have set up the bait and armed themselves with their rifles and night-vision binoculars and monoculars, it's already approaching midnight. The furious office workers who decompress at the squash courts and sauna rooms at the sports centre by the golf course have all gone home. Apart from the distant burr of A-road traffic, the night is silent. It's hard to imagine that this former landfill site, now transformed into lunar pastoralia, has become overrun with foxes, or that the foxes are so cocky they have been emerging from the long grass to run after and pinch golf balls that have landed near them.

Bruce drives the Land Rover with his left hand while holding a flash light with his right hand. It's a slow and painful way to navigate the darkness; as night goes on, bolts of carpal-tunnel pain shoot up his arm. He doesn't have a map, and the course is so bumpy that the car nearly topples on to its side many times. Some nights, he recalls, they have fallen into marshy ponds and required tractors to pull them out: "Funny thing was that there were no frogs in the ponds; the foxes had eaten them all."

We keep to the perimeter of the course. The higher skyline vantage points here give us a sense of brief mastery, as if we are African game-rangers looking out across imperial plains. Every few seconds, Bruce blows into his rabbit squealer, an audio-contraption that emits a fake come-on, an insidious serenade designed to encourage wallflower foxes to emerge from their hide-outs, only for them to be felled seconds later by a bullet to the head. But tonight, the foxes don't seem to be responding. Many, Bruce surmises, won't even be on the links; they'll be wandering nearby streets foraging in the bins by bus shelters and the backs of local shops.

Suddenly, amid the blackness, I see two tiny specks of light. Before I can work out what it is or how far away it is, there's a rifle snap. "Did you see that one?" Bruce asks excitedly. "I love it!" Then, pointing to congested trees he estimates are four hundred yards away, he tells his marksmen: "There's three there." A couple of the foxes, perhaps naïve about the risks they face by showing their heads, perhaps starved of food for a week and getting a noseful of the rabbit scent, are moving towards the car. "Take it carefully, Julian," says Bruce. "Lots of time. Let him come off that track. Here he comes." One shot – and the fox is dead. A couple of seconds later, and another one is dead.

The speed of the shootings disorientates me. After spending so long looking for signs of animal life, it seems wrong, or at least peremptory, to put an end to that motion. I feel complicit, as if I'm taking part in a snuff movie. Bruce senses my unease: "The bullet travels at 4200 feet per second. That's three times the speed of sound. They don't even twitch. They're stone dead before they know it. If you hit them in the head, it goes straight through the spinal column. Sometimes it blows them virtually in half."

After a while, we drive to where the dead foxes lie. Bruce and his men, after checking to see if the slain animals are cubs, vixens or dog foxes, do a beauty audit. Some, the steam still coming off their warm skin, are elegant and pretty: "He's got a fair old mask on him." "She's a big fat vixen; got a lovely skin that one." Others, those with poor diets, scabby flesh, and mange creeping up their tails, are not. Age is assessed by peering at their teeth; town foxes, who normally survive only a couple of years, usually have a good set. Those who also have good furs will be skinned. As recently as the early 1980s there used to be 76 tanneries and furriers in London, mostly around Hackney, where fox-skin coats were tanned, made up and sold. Most have disappeared. These days, a decent coat sells at game fairs and country shows for £25 to £30.

Tonight Bruce kills twelve foxes. It's a tidy haul. Enough to suggest that he has a better knowledge of the urban-fox population across London than his scientist critics. As the hours tick by, and repeated circuits of the course yield no more fresh kill, Bruce talks about why he loves his job. Partly, it's an issue of aesthetics. He recalls the unusual shrubs and flowers he has spotted on his nights across London. Changing weather patterns have brought greater varieties of butterflies and moths to the capital. The elephant-hawk moths he saw on the banks of the Thames near Silvertown probably came up on boats from Europe. The technology that he uses allows him to see colours and patterns invisible to everyone else: "You get your own laser show just through looking through the infra-red illuminator. The moths and flying insects look extraordinary. You gaze up at the sky on a clear night and you can see thousands more stars than you could see with the naked eye. It really throws them back at you."

Soon, he starts to rhapsodise about the foxes: "I don't do it for the sport. They're cunning and they're crafty, and they're also very stupid at times, but they have a fascination and glamour. I think they're absolutely lovely, much more nice-looking than the grey squirrels people go on about. Their grace, their ability to adapt to everything, how beautiful they are – they're my favourite animal."

XI

Elliptical Night:

Sleep Technicians

ELLIPTICAL NIGHT:

SLEEP TECHNICIANS

There exist Londoners who do not wish to spend their nights partying or raising their hands in the air at the behest of a sweaty, badly-shaven man nodding his head behind a pair of turntables. These Londoners only desire is to sleep. But, for reasons that they can't quite gauge, their bodies and minds won't let them. They have been barred from sleep. They feel jammed, frozen. Trapped in the glare of an endless midnight sun.

And so, each weekday evening, a select handful of them will head towards a hospital near Westminster Bridge where they will be masked, wired, strapped up and lain down in rooms that insulate them from whatever is going on in the city. For the next few hours, up to and beyond dawn, every snore, spasm and twitch of their eyes will be monitored by men and women with a very unusual job: they are sleep technicians.

"We're a nocturnal gang. We're a special unit, a minority unit of people who not only work odd hours, but beat the rat race. Think about it: every day you have 22 million people commuting to London, the nerve centre of the UK, as well as the 8 million who already live here. Even with those numbers, we see and hear things that no one else does. We're guardians of the unknown."

Londoners today get 25% less sleep each night than they did when H. V. Morton was out noctambulating. Fewer and fewer workers are allowed to leave their work behind at the office: they are foisted with mobile phones, Blackberries, wi-fi connections – technologies that allow them to be linked and live-wired to the demands of their employers. They are held in a digital lasso, forced to dispense with the idea of down time, and spending late-night hours pinging draft spreadsheets to equally bleary-eyed colleagues. Sleep, according to the rules of the post-industrial game, is for losers, for those who can't hack it in the upper echelons of the new information economy.

Capitalism detests sleep because it's a time when nothing gets sold or consumed. It used to be that television and radio stations shut down at midnight. Now hundreds of cable channels exist to peddle gaudy trinkets, soft rock box-sets and re-runs of 1970s cop shows. Disc jockeys, whether smooth-tongued presenters of mellow gold or hyped-up jabberers on the pirates, boast that they are up all night. The flats of Londoners have become entertainments arcades full of games consoles, sound systems and ever-growing flat screens and monitors: the party, whether or not there is anyone else to share it with, need never stop. The shift from dial-up to broadband means they can inexpensively while away the darkness chatting and surfing and paying bills. The computers are always on; and so, it seems, are growing numbers of Londoners.

Super-stimulated by the buzz of entertainment and the pressure of work deadlines, Londoners, to a far greater degree than people elsewhere in the country, are falling prey to sleep sicknesses. Their body clocks are all over the place; they skip key meals, but pop down the local 24-7 to stock up on coffee or cheap bars of chocolate. Maybe they douse themselves with a glass or three of Shiraz. Drugged up, overdosing on excess, thrusting their brains and bodies into conflict, Londoners are hitting epidemic levels of obesity and courting diabetes as never before. Shift workers, forced to do 8–12-hour stints at irregular intervals, are most likely to get ulcers or heart diseases.

Sleep technicians are sure that poor diet is one of the key reasons why London is unable to sleep. They calculate that over a third of its population suffers from insomnia, and that it interferes with the working lives of ten per cent. Yet they say it's a topic that they study for only nine minutes as part of their four-year-long medical educations. They also say that sleeplessness is a scandal politicians do not wish to broach, not only because dealing with all its victims would bankrupt the NHS, but because the kinds of changes needed to combat the problem would require a social and economic revolution.

"I haven't been able to sleep for so long. Every night I'm still up and it's like someone's hit me over the head. My mind's doing a 100-metre dash, someone's removed the finish line and it just keeps going and going and going until you literally feel like your head's a cage and there are mice running around in it. You can feel them with their claws burrowing at your skull further and further until as if they're going to emerge. You feel the physical pain in your head because your mind's just racing racing and you just cannot shut it off."

Nights begin at around 7.30pm. That's when the patients, from kids whose snoring can be heard at a distance of thirty feet to men in their seventies, start to arrive. Many suffer from apnoea, a condition that stops people from breathing for periods of up to a couple of minutes and makes them explode in volcanic gasps during which they try to get some oxygen. Though they often complain of feeling sleepy – understandable since they may wake up 800 times a night – it's common for them not to know anything about their maladies; it is their long-suffering spouses who haul them along to the clinics.

After all the consent forms and allergy questionnaires have been completed, the patients get into their bed clothes and then sit for up to an hour while they are prepared for their sleep tests. Silver nodes and a series of wires are attached to different parts of their heads, faces, fingers and bodies using conductive paste that allows technicians to check brain activity and whether certain muscles are overactive. It can be a tricky procedure: some middle-aged men and fashion princesses turn out to have hairpieces or weaves stitched into their scalp wigs making it hard for the techs to fasten the wires.

Some of the patients get excited by having to don electrical wardrobes. They take mobile-phone pictures of themselves which they send to friends across the city, and are thrilled by the idea that, just like *Big Brother* contestants, there will be cameras filming them throughout the night. Other patients are more nervy and diffident. They regard the idea of being watched as they sleep as voyeuristic, a breach of the privacy that the darkness of night time is meant to usher in. They feel naked without the protective armour that cosmetics provide. They worry about being caught dribbling or reciting names it would be politic to keep to themselves.

More than that, they have become used to sleeping badly, and worry that they will have to adopt new regimes. They fear the electrodes are miniature cattle prods that will shock or electrocute them. As they get hooked up they confide in the technicians about how they have not had sex with their partner for two decades, or how they once became so bleary and depressed they tried to take their lives.

The technicians have not been trained as marriage counsellors or as psychologists, but they soon develop a strong awareness of how vulnerable the sleepless can feel, even when shown the equipment that might improve their symptoms. It is common for patients to panic upon learning that they will have to wear a respiratory mask as part of their examination. Younger ones, who suspect the tests will determine they have to wear those masks all their lives, cry out of vanity; they fear no one will ever kiss them again, or that their partners will regard them as life-support-machine victims. Even older patients claim they will feel muzzled, claustrophobic; a fireman, whose job forced him to wear a mask all day, began to hyperventilate at the prospect of being attached to another one all night, so much so that he ran out of the hospital.

"People have strange ideas about what we can and cannot see. There was one time a girl came in to help out her mother who didn't speak much English. But she was pretty old fashioned herself. I explained to them both about the cameras and that I'd be using them to see exactly what was going on in the mother's brain while she tried to sleep. The daughter totally misunderstood: she thought we could read and work out exactly what her mother was dreaming. She thought the images would appear on the TV screen. She started warning her mother: 'I know you don't like my husband Jorge. But you'd better not be saying bad things about him, or I'll be really mad at you.'"

Sleep technicians are security guards who watch over their patients' physical and abstracted selves. They use their eyes and infra-red cameras to monitor pulses, the flow of breath, rates of desaturation, all of which give indications of long-term breathing and sleeping problems. The screens are high-powered enough to allow them to pop to the toilet and on their return rewind the action just like TiVo. They'll spot tics and twitches, violent teeth grinders, kicky leg shakers who can't contain their restlessness. The effect, especially when they notice men playing with themselves, can be peepshow-prurient.

The apnoea sufferers, especially those with a history of heart disease, are regarded as the worst. Their airways collapse. They stop breathing for long intervals, are totally still, and then erupt; they are, in effect, fighting for their lives every couple of minutes. Their heads may keep shaking, and they often raise their arms in the air, like drowning men out at sea waving for attention. It is common for younger technicians, in spite of their training, to panic and assume that the patients are in the middle of a sustained seizure. But even more experienced staff can be upset by the violent noises and their own helplessness. They feel morbid, mute witnesses to a dark, disturbing sorrow.

By 2 am, most of the patients have drifted off. They can look like overgrown babies, some anguished, others wreathed in subtly beatific smiles. Most of them though, breathing masks covering their faces and all muscle tone having disappeared, resemble bodies laid out in a mortuary. The technicians who watch them from adjacent rooms stare at them sometimes, trying to gauge their characters from their obscured physiognomies and dormant postures, but it's impossible; the images on screen are transmissions from another galaxy. Even the research scientists in less well-funded units, who spend nights in the same room as their patients, admit that the men and women sleeping a couple of yards from them, behind a drawn curtain, may as well be miles away.

These technicians betray their own beliefs every time they come to work. They know full well, and they tell their patients as much, that it's sensible not to work long shifts. That regular breaks are vital for sustained good health. Yet they themselves routinely work for twelve hours. By 3 am, they need artificial highs to stay alert. The Krispy Kremes come out. They start swigging sports drinks or listening to techno on their i-Pods. Graduate researchers who tend to work with less afflicted sleepers, will lie on the floor, using small torches to read journal articles or comic novels, all the time conscious that they mustn't laugh out aloud.

"You could say the patient, to us, is a series of lines and shapes on a screen. We spend the night reading polysomnographs. We read waves, not books. We're checking the frequency and depth of brainwaves. They all have a different shape and character. Clear alpha waves look gorgeous, like lovely big fat rolling hills. Then you have delta waves. Then if there's something going wrong you see on the monitor these violent spikes: awkkkk!"

Sleep is a cave, an Anderson shelter to which city dwellers can retreat from the stresses and strains of metropolitan life. It's a departure lounge, where men and women can use dreams to rewind and choreograph afresh dismal past events, and where they can fly away to strange new zones more brilliant and polymorphous than those with which they have to put up normally. Without sleep, and without dreams, they feel trapped in the middle of a violent, concussive city. And yet, they also feel invisible, insubstantial. They want to apologise to friends for not really being present. They lose their train of thought easily. Sentences fizzle out. The whole day becomes a series of stuttering, apologetic ellipses.

Insomnia is smog, perpetual drizzle. Sufferers feel foggy all the time, as if they have scarfed down a bottle of Prozac. Stupid too: how can an activity so simple and basic become such a problem? They study and restudy their diaries, paranoid self-scrutineers berating themselves about the double espresso they ordered at a restaurant the evening before and how it must be the reason they only got two hours of shut-eye. They begrudge sleeping partners their restfulness for it compounds their isolation and unhappiness. Radio becomes a fix, a late-hours accomplice. Night starts to take over day: they think and stress about it when they should be working. They become afraid of it, hoping and worrying that the next evening will finally be the one they manage to enjoy some quality sleep. They worry that they might be in no fit state to deliver their boardroom presentation the next morning. All the time they feel broken, in pieces.

Sleep, for insomniacs, is a wan, portable concept. They catch it while they can, whether it's a couple of minutes at the work's canteen while their pals are discussing Leyton Orient's recent dire form, or slumped over the free papers on the tube ride back home. But it's never remotely enough for them to stop envying their more functional friends for whom the night is an orderly, sequential experience: get back to the flat, have dinner, clean up, watch telly for an hour, go to bed, fall asleep by 11. "At 2am," says a care worker, "I find myself looking down a long corridor that stretches forward to the morning. It is lined with a bed, sheets, pyjamas, blanket, clock, gown. But I find that the spaces between all of these things keep opening up and extending. I can't navigate through."

For insomniacs, night has no end and the day no beginning: looking at the clock merely compounds their suffering. The beds they climb into irritate and annoy: sheets feel rough, pillows too lumpy, even the night clothes are prickly. They lie on their sides, facing the ceiling, at oblique angles; but they can never find a comfortable spot. They feel lonely and homeless, that there is no place in the world that they can treat as a safe haven. So they drift, always gauzy, often walking into doors or ending up with scratches or bruises on their legs without any idea of how they got there: "You don't have any barrier or shield to protect you. It's as if all your edges are permeable. If someone were to draw you it would be – a wavy line."

Insomnia corrodes memory. The sleepless can't remember if they locked the back door or fed the cat. They're not sure which afternoon they're meant to be attending their children's sports day. None of their experiences are processed or filed properly. They become trapped in the present, unable to mature or evolve. They feel insensate, estranged: all of London, its adventures and incremental wisdoms, is negated. It becomes a faded photo.

Sleeplessness renders the bedroom a holding cell, a deprivation chamber. Some insomniacs, fearing that the Heart DJ will play *Come Away With Me* for the fifth night in succession, flee outdoors. The streets of London are populated with men and women who are not going anywhere, but wraith around trying to kill time until they feel capable of falling asleep. They describe themselves as the undead, zombies in overcoats. Amidst the crowds of Soho and Tottenham

Court Road they stand; pieces of blotting paper that absorb the colour and noise around them. They are part of the city and standing outside of it, immersed yet detached. They look up at the sky to make a wish. They stare into the distance. But there is no dark at the end of the tunnel.

"I used to think people who slept were bourgeois and conservative," admits a lawyer. "I saw them as belonging to the bovine, complacent herd of Londoners I despised. I thought life in London was a competition, and that not sleeping gave me a competitive edge. I don't so much now. Still, I wouldn't say that being an insomniac makes you a more caring person. When I'm wandering around King's Cross at night, I occasionally see people who look like me and I just scurry away. I worry we might compound each other's sleeplessness."

Insomniacs inhabit a strange, hushed universe in which even the thoughts that go through their mind seem a jangling transgression of night's silence. Increasingly, though, and whether or not they traipse the streets alone, they know that London is a persistent noise field. Whether or not they open their windows at 3.30 to let some air in, they're sensitive to every horn-honk, avian-police helicopter and knot of hooded youths cussing and laughing. They hear the illegal construction crews of east European labourers getting ready to work. They hear brewing company workers rolling steel barrels of ale across the pavement to local pubs. They swear that even the birds are tweeting and crowing noisier and earlier these days.

"What's good about not being able to sleep? Well, I know a lot of facts about inorganic chemistry from all the Open University programmes I've watched. I know how to ask directions of a stranger in lots of European languages. Mainly I've learned how crabby and deprived insomnia leaves you feeling. I can see the Thames from my flat. Some nights I feel like I'm

**drowning. I want to rush out and climb St Paul's or hang off
the London Eye. But I never even leave the front door. My life,
my dreams: they've become totally dimmed."**

Sleep technicians are often annoyed – by patients who keep asking
if they can get up to go to the toilet, when some of the wires fall off
at 5 am invalidating the whole night's tests – but they are rarely
scared. Those that work in the older hospitals, places widely
believed by both medics and administrators to be haunted by
ghosts, recall hearing bumping noises or creaking staircases. But
the most scary moments, more so than the patients' erratic breathing patterns which are ultimately physiological, are the night-
mares. Patients will scream and yell at the tops of their voices.
They will make violent threats and start to rain down curses on
family members. The technicians are initially fascinated and eager
to hear more. Soon though, they start hearing fragmented re-enact-
ments about the sleeper being molested by, or molesting, someone
close to them. They back off quickly, and try to focus on the
polysomnographs.

Some sleepers even suffer from night terrors, traumatic occur-
rences usually the preserve of pre-teens who feel they are being
attacked by amorphous black shapes in the corner of the room.
During these terrors, the victim, whose eyes will be fully open
though they are actually asleep, will scream hysterically for up
to half an hour. One technician was monitoring a patient who had
five terror attacks in one night. He ripped off his mask and flung it
across the room. Then he started leaping on the side of the bed
where he though his wife was lying, in order to protect her from the
creeping blackness. All the while he sobbed uncontrollably.

**"This is a rational universe, right? And London's a modern city,
no? Well, I've seen kids not even drawing breath they're flail-
ing and shaking and screaming so hard. You think: their vocal**

chords can't take it. It's like the devil is in the child. You want to comfort them and wake them, but that might make it worse. You're their carer and you're powerless. They're totally non-responsive. I know of so many parents who don't want the neighbours getting upset; they strap their screaming kids in the back of their car and go driving around London. Round and round and round – till the rhythm drugs the child."

Early morning in a sleep laboratory has its own smell. Every technician, whether giggling or groaning, will admit as much. In summer time, the machines can give off up to 40-degree heat. The odour of methane, plus the fetid breath when patients' masks are removed from their faces, makes for a distinctive perfume. Windows and doors are opened hurriedly.

The patients themselves tend to be bewildered. Decades of fragmented sleep have left them with no knowledge of what it means to get a good night's rest; they take their grogginess and irascibility for granted. Now they look around, shorn of the uniforms or fancy threads that give them status during the working day, suddenly becalmed and beaming. They stare at the technicians as if they are miracle-agents who have brought them back to life, wrought Lazarus-like transformations.

When told that the only remedy for their condition is to wear breathing masks every night for the rest of their lives many are upset and dejected. Others, especially older patients, don't care: "It's not as if me and the missus are having sex all the time anyway." They calculate it's better and more effective than sleeping pills. A lorry driver says: "I used to feel that my soul had been taken away from me. I felt shallow and empty. I knew there was something missing. I'd love to spend the next few years getting back all the sleep I've missed in my life."

"Being a sleep technician is rewarding, but it's damn hard work. By the morning I'm knackered and a moody bastard. I see all the commuters coming into London with their bloodshot eyes and their fat necks; I know they're not breathing or sleeping well. You think to yourself: half the people in this city are ghosts. Still, I go home through Hyde Park; sometimes, no matter how shit I feel, I see the mist on the ground and the steam rising from the lakes and I can't help but smile. I always think: the city's waking up. The city's given a bit of soul back to me."

XII

Midnight Pilgrims:

The Nuns of Tyburn

MIDNIGHT PILGRIMS:

THE NUNS OF TYBURN

There has been a woman on her knees at the top of Bayswater Road for over a hundred years. Not the same woman: she may be young, a former nurse from Brisbane; she may also be a Scottish ex-teacher in her sixties, whose advanced age is only betrayed by her unsteadiness when she gets to her feet. Both of them are nuns at Tyburn Convent, a Benedictine order that requires its sisters to wake up during darkness and kneel before God as part of a ceremony called the Night Adoration. They pray for the souls of Londoners – many of whom will be frantically pursuing narcotic communion or surrendering themselves to a kind of ecstasy about which the sisters will not have heard.

The Convent has a bloody history. It stands near the site of the King's Gallows where, from 1196 to 1783, thousands of criminals were hung, drawn and quartered. Between 1535 and 1681, 105 Catholic martyrs were executed here too, and fragments of their skeletons as well as segments of noose-rope are on display in the crypt. One night in June 1944, a German doodlebug struck the convent, wrecking the chapel and cloister. The sister who was in the middle of her Night Adoration shift ignored the falling masonry, as well as the unusual sight of nuns without glasses or false teeth crawling out from their wrecked cells, and carried on regardless. Later, an American GI who had been stationed across the road, in a military camp at Hyde Park, helped attend to her shard- and splinter-lacerated face.

The Tyburn nuns survived that night and even the mischief of mice that emerged from behind the skirting boards. They survived not just the war, but everything that has tested or buffeted them since: the prostitutes who used to line the railings in the streets outside, the tramps who abused their hospitality to rob the premises, disability regulations that forced them to instill elevators they could ill afford. Their convent stands almost at the heart of London – at the bottom of Oxford Street, a screaming retail Mecca clogged with chuggers, megaphone-sporting evangelists and knock-off perfumiers.

Yet the sisters occupy a kingdom far removed from the swish of traffic and the blare of commerce. Each is an enclosed contemplative who has taken a vow of poverty, and who spends part of the day cooking plain fare and sewing clothes. They have no television, radio or daily newspapers. Their library is small, and with the exception of a few volumes of Shakespeare and Dickens, contains mainly historical or geographical books. During the Feast of Epiphany they each pick a country whose history, local saints and ecclesiastical traditions they research before holding a small exhibition and teaching other sisters about it. They also direct their prayer towards that country.

O London, take pleasure in your people:
Beautify the meek with salvation

The Tyburn nuns have been based in London since 1903, but their knowledge of London is minimal. They rarely go out, and are allowed to meet their parents only once a month. Dressed in floor-length habits, and scrubbed free of cosmetics, they claim to be married to Christ. Their lives are purely devotional, dedicated to the constant adoration of the Blessed Sacrament. Each day they are only allowed one hour of conversation, a period in which they often play snooker or volleyball; during the remaining hours they sing Mass seven times. A shift system operates at night so that, every hour from 8.30 to 5.30, one of the nuns will be engaged in committed prayer for sixty minutes.

"We quite accept that this life is not for everyone," says one of the sisters. "Poverty is hard; for those used to bringing in trays of McDonald's to eat at home at night, it's difficult to understand the joy of cooking or of quiet eating. The workmen who come here often don't believe the difference or even that they're in London. But I don't think it is right to say that our lives are defined by an absence of noise. The silence is not empty. It is rich and replete. The silence is alive with the presence of God."

The nuns speak, whenever possible, in the language of affirmation. They do not like to carp or criticize. But, like many of the street-sweepers and bin-men who keep clean the city beyond their vestibule, they cannot help but see contemporary London as a place of excess and satiety. It is so plumped and cushioned that its inhabitants are insulated from the necessary rawness and proximity needed properly to access God. It is decked out with gaudy baubles, unChristian light. The nuns, by contrast, do not see regard discipline as a shackle or as an affront to their imaginations; rather, it's a liberation. It frees them from over-consumption and licentiousness. "Our Convent," they laugh, "is better than any soap opera. We learn about ourselves and each other. We watch the sisters."

And yet the nuns do not think of themselves as entirely segregated from the city. News, especially of a religious nature – a Royal wedding, the death of a Pope – always reaches them eventually. Clippings from the *Catholic Times* are often circulated. A website exists to promote the convent's work and to encourage financial donations; it also features footage of individual sisters giving advice to potential new recruits and explaining the Rule of Benedict – albeit introduced with what seems like deliberately prurient language: "Watch our nervous, camera-shy nuns reply via video footage."

Mostly, the sisters communicate with London through and in the act of prayer. Londoners often ring in to request that the sisters pray to God on their behalf. A young woman gets in touch because her mother is having an operation the next morning. An anxious young boy cries that if West Ham lose their home game at the weekend they will be relegated. A community gardener is concerned that she can't contact her cousin who lives on an Asian island that has been hit by a hurricane.

Details of each call are taken down and pinned on a noticeboard for the sisters doing the Night Adoration that evening to re-narrate to God. They become not just penitents, not just humble zealots doing God's work, but channels for the city's deepest fears and yearnings. They traffic between the secular and the divine, reassuring both believers and non-believers that there is a grander, benign surveillance system monitoring them than the CCTV cameras at the end of their street. To their callers, especially those that used to sneer at or be sceptical about religion, everything that was particularly odd and anachronistic about the Tyburn nuns – their quietness, their commitment to routine and repetition, the different philosophy of time to which they cleave – now becomes a strength, a sign that they have an alternative set of emotional and spiritual resources that might succour them during their hours of need.

"I have heard it said that men and women in London struggle to pray in absolute silence. It is a very noisy city. When I pray I have to empty myself of all my inabilities and my weaknesses so that I may fully occupy that place of silence and stillness in order to receive the Lord. Ultimately, I think I hear the voice of God: it is in, from and out of the silence. It is in, from and out of the darkness of London at night time."

London, my London, why have you forsaken me?
Why are you so far from helping me,
And from the words of my roaring?
O London, I cry in the daytime, but you do not hear me.
I cry in the night season, and am not silent.

H. V. Morton thought that night time was the right time "to make promises to London, to pray to London, to plead with London." He rhapsodized about the silhouette figures cast by St Paul's Cathedral and smaller churches at night, but said nothing about the people who worked or prayed in those places. Religion, for him, was an issue primarily of architectural or historical importance; an occasion both to celebrate the aesthetic qualities of houses of worship, and to unfurl century-spanning narratives about the role of Christianity in shaping and underwriting the best of England. The functions that religion served, however, the pain it staunches and the emptiness it struggles to fill, was territory too dark for him to do anything more than gesture towards.

That pain and that emptiness is something that Malcolm Hunter deals with every day and every night. He is a friar at St Michael's Church in Camden Town, an area known for its high rate of home-lessness, female suicides and drink-related deaths. In the last couple of years, a ragged army of junkies, pimps and homeless men and women, evicted from the back streets of King's Cross in order to reassure incoming gentrifiers and homesteaders, has been moving here. The number of tourists heading for the outdoor market stalls and of lank-haired students dancing at grindie clubs gives the area a feel-good energy. "It's not real," argues Hunter. "What we at this church do is peer below the scabs and the surface of daily London. We unpeel all the protective layers."

Hunter grew up in Kenya and still misses its expansive terrain and uplifting light; in London everything and everyone seems knot-ted, tight, tough. At time he thinks of it as a sewer. All around him is weakness, addiction, fake stimulation, social inequality. He's in the middle of it, barely able to keep himself together enough to protect the people about whose welfare he anguishes. "I try to see God in that person who's in the City, who walks by and ignores me. But there are more attributes of God in the person of the asylum seeker;

that asylum seeker has known what it's like to have lost everything, and to have lost loved ones. For me, it's the weak who have the greatest relationship with God. The modern city doesn't want to acknowledge that though; it says, 'How can you be strong when you're crying your eyes out?'"

Faith makes him see London differently from the non-believers do. God is everywhere, permeating day and night every inch of the distended city. He is present, at once potential and actual, in everything about the capital that is good or bad, beautiful or ugly, fleeting or persistent. "When I see drugged young girls outside my church injecting themselves, or struggling to find a vein, or even missing a vein, well, I don't know what other people see. But, for myself, I see Christ. This is a manifestation of God. All people are in Christ. There's actually God at the end of that bloody syringe needle."

> **I am poor and needy,**
> **And my heart is wounded within me.**
> **I am gone like the shadow when it declines...**

For Hunter, prayer on its own is not enough. Faith has to be expressed via good works in the community. For some time he had been keen to create a night shelter for the homeless and vulnerable. Funds being scarce, he hatched a plan to spend ten nights on the roof of St. Michael's, sleeping without blankets or duvets and fully exposed to the sky in order to reach his target of £100,000. Van drivers honked their support, local schoolkids winched up cakes to him, and reporters from television and the press alighted on that church from all around the world. It was a great story, a trendy vicar engaged in a spot of reality-theology for the masses, a religious version of a David Blaine spectacle.

The response from some sections of the church establishment was rather different. They were intimidated by the media spectacle and suspected vainglory on Hunter's part. Worse, they implied that his newly-acquired aerial view of the city verged on blasphemy. It was seen as an attempt to mimic and rival the Lord's Archimedian perspective. But such suspicion and gossip was as nothing to the dark vision of London that was revealed to Hunter during his week and a half on the roof. Each night he stared in sorrow as the city's downtrodden and rejected – the self-harming ex-soldiers, the scroungers and pickers, the sozzled drifters – pissed on the streets, faced off against each other with knives and broken bottles, jacked up on nearby rooftops. The enfeebled movement of the weak rather than the purposeful mobility of the successful. Being perched above the city brought Hunter no comfort or reprieve from earthly worries; all he saw were broken individuals struggling to battle their demons. It was summer, the weather was lovely, yet a fog of fear – his and theirs – hung over London.

That being stuck alone on the roof would be uncomfortable was to be expected. However, the Bishop of London, when he met Hunter, voiced another concern: "Can I ask you that you don't run into yourself?" "Actually what happened up there was that I ran into God in a far deeper way than I had ever expected to. But what really hurt me was that I didn't see Christ's resurrection. I only saw his crucifixion. I could see only pain. I mean, it was an amazing view; I had freedom, a bird's eye view of the city, I could see everything. Canary Wharf, the financial City, massive suburbs at St John's Wood and Hampstead where some of the wealthiest people in the world live. But I also saw the Royal Free Hospital, the men's hostel at Arlington House: so much isolation and loneliness. You think: it's all a load of bollocks. London's not alright. It's an absolute charade. And though my prayers were and are full of joy, I also get really angry and sad and tearful and despondent and full of despair."

From the east, and from the west,
From the north, and from the south,
They wandered in the wilderness in a solitary way:
They found no city to dwell in.
Hungry and thirsty,
Their souls fainted in them.

All across London churches are disappearing. As its population gets younger, congregations disperse and decline. The buildings, like those of Victorian schools or Jewish soup kitchens, are snapped up by property speculators intending to repackage them as high-brow heritage for the discerning flat-buyer, pre-packed moral ballast for the floating consumer. Spirituality becomes relegated to the past, and the buildings that anchored local neighbourhoods and gave their residents a hallowed third space between work and home are transformed into fuzzily nostalgic carcasses.

But if old churches are disappearing, new ones are springing up all the time. The Poles and Lithuanians and Brazilians who have migrated to London revive formerly moribund places of worship or establish new ones. They, like Pentecostalist groups from Ghana and Nigeria, or Muslims from Iraq and Pakistan, band together to create new gatherings in rundown premises on the edge of the city, in borrowed industrial-estate offices, in the living rooms of their council flats. This is start-up divinity, enterprise theology from people who have embarked upon a journey whose end they cannot imagine and whose success they cannot forecast. Religion is part of their survival kit as they go about trying to plant a stake in this new city in this newfoundland.

"In the last few months since I lost my job – well, it's been hard. My mammy died last year. I don't have a boyfriend or a husband. I got this part-time job in a friend's shop. It don't pay too well. I don't rarely go out these days. Every night when I'm

walking home, I see all the market leftovers and the rubbish and the glass. I just think: well, that's me. So when I pray, I'm just praying to God that things will get better soon, and that I won't always feel like the leftovers and the rubbish."

One Friday every month, in the over-lit and over-heated refectory of a Comfort Inn on the South Lambeth Road, thirty to forty Londoners of Nigerian descent gather to take part in a special late-night-worship service. They are members of the Kingdom of Life Prophetic Church, a Pentecostal group led by Bishop Joel Oluwafemi. A sharply suited man who racks up a lot of air-miles, he appears regularly on a Christian cable network broadcast from Acton called Wonderful TV, and is also responsible for a growing number of books and DVDs. Tonight, because he has to catch a flight at 6am, the service will only last from around 10pm to 2.30am; on other nights, it has gone on much longer.

The atmosphere is very different from that at Marble Arch. In place of stillness, there is swaying, arm-flinging, freestyle movement across the refectory. In place of silence, there are whoops and cheers, calls and responses, the sound of women gently hitting their heads against walls. They rock to and from, swaying from side to side and begging contrition of God. The chanting is simple and endlessly repeated: "We should rejoice"; "In the name of Jesus, I worship you for what you are"; "God will give us, God will give us." Unnoticed, a hotel porter pops in to get a coffee from the automatic dispenser.

The congregation, many of them women, many also middle-aged, have come from all across London to be here. Some have brought their children, mildly sulky young teenagers in their Friday-night best of baggy jeans and white rollnecks, to join the celebrations; as the evening goes on, and the sermons give way to more extemporized rhythms and ecstasies, their scowls turn to smiles, and their hips start to roll. The atmosphere, even when it deals with shame and mortification, is affirming, polyphonic. Through a window I can see a double-decker bus parked outside sports a huge poster advertising a new series of *Lost*. But the people inside, midnight pilgrims searching for reassurance and joy, have found, if only briefly, what they were looking for.

The Bishop starts to deliver his sermon at 1 am. He explains that it will have to be a short one; it lasts well over an hour. His rhetoric, pitched somewhere between that of a travelling salesman and a high-school teacher trying to ensure that even his slowest students have grasped the fundamentals of life science, is spruced up with metaphors mostly concerning fast cars. We are in London, he tells us, and yet we must wage war against the lassitude and looseness with which London abounds. We may be children of God, but we should think of ourselves as soldiers. What we need is vision – and focus. To do this, he exhorts the nodding congregation, it is not enough to dwell on inequalities, or on social division, or on racial slights; rather, discipline must be embraced – no street fashions, no corn-row hairstyles, no slack body language. Nothing that might create anxiety in a potential employer.

His words go down well – especially with the mothers. Soon, the music and the melisma strike up again. "Vision and focus." "God will hold us." One worshipper squeezes my arm: "The longer the contemplation, the cleaner the light." These melodies, however babbling they may seem to outsiders, and however exiguous they emerge from the mouths of the more elderly people here tonight, will sustain them over the next month. They're still singing them as they leave the Comfort Inn, the melodies, strained and thin, floating up into the night sky and drifting across all of London.

I am so troubled I cannot speak.
I have considered the days of old.
The years of ancient times,
I call to remembrance my song in the night....

Listen carefully. People are praying tonight. The blue-light ambulance driver tearing through the streets of South London in the hope that he can still deliver a hit-and-run victim to A&E before it's too late. The young Chinese vendor who has spent the last few hours ducking in and out of New Cross pubs trying to sell knock-off DVDs, and who now sees a group of toughs looking enviously at his backpack as they move towards him. A gap-toothed teenager shivering in a back street off Bethnal Green Road hoping for a john to blow otherwise her boyfriend-pimp is going to start whacking her again. The shivering Iraqi who has been standing in line outside Lunar House since 4 am so that his claim for asylum may be heard as soon as possible.

Listen carefully. There are people praying tonight. The barger who heard someone scream, followed by the splash of water. The exorcist battling demons and devils. The insomniac gargling down half a bottle of sleeping pills. The graffer trapped between the torch-light of security guards and an oncoming train. The Samaritan who hopes the thud he heard on the other end of the line doesn't mean what he thinks it means. The Botswanan cleaner worried her boss is going to fire her for being late again even though that was because some guy had flung himself onto the line at Willesden Station.

Prayer is the true language of the night. It is the sound of London's heart beating. The sound of individuals walking alone in the dark.

(Place of publication is London unless otherwise noted)

Almqvist, Björn and Emil Hagelin, *Writers United*
    (Årsta: Dokument Förlag, 2005)

Alvarez, Al.: *Night: Night Life, Night Language, Sleep, and Dreams*
    (New York: W.W. Norton, 1995)

Bartholomew, Michael: *In Search of H.V. Morton* (Methuen, 2004)

Berger, John and Jean Mohr: *A Seventh Man* (Penguin, 1975)

Brassai, *The Secret Paris of the 30s* (Thames and Hudson, 1976)

Burke, Thomas: *Nights In London* (New York: Henry Holt, 1916)

Burke, Thomas: *Limehouse Nights* (Grant Richards, 1916)

Burt, Jonathan: *Rat* (Reaktion, 2006)

Calle, Sophie: *Exquisite Pain* (Thames and Hudson, 2004)

Chatterton, Paul and Robert Hollands: *Urban Nightscapes: Youth Cultures,*
    Pleasure, Spaces and Corporate Power (London: Routledge, 2003)

Dunkell, Samuel: *Sleep Positions: The Night Language of the Body*
    (New York: William Morrow, 1977)

Ekirch, A. Roger: *At Day's Close: A History of Nighttime*
    (New York, W.W. Norton, 2005)

Evans, Christopher: *Landscapes of the Night: How and Why We Dream*
    (New York: The Viking Press, 1983)

Fletcher, Geoffrey: *Down Among The Meths Men* (Hutchinson, 1966)

Gastman, Roger and Darin Rowland and Ian Sattler: *Freight Train Graffiti*
    (Thames and Hudson, 2006)

Hujar, Peter: *Night* (San Francisco: Fraenkel Gallery, 2005)

Jerrold, Blanchard and Gustave Dore: *London: A Pilgrimage*
    (Grant and Co., 1872)

Katchor, Ben: Julius Knipl, *Real Estate Photographer: The Beauty Supply District*
    (New York: Pantheon, 2000)

Maddox, Adrian: *Classic Cafes* (Black Dog, 2004)

Marvin, Simon and Will Medd: 'Metabolisms of Oboe-City: Flows of Fat
    Through Bodies, Cities and Sewers', in Nic Heynen, Maria Kaika and Erik
    Swyngedouw, *In The Nature of Cities: Urban Political Ecology and the Politics of*
    Urban Metabolism (London: Routledge, 2006)

Matthews, Anne: *Wild Nights: Nature Returns to the City* (New York:
    North Point Press, 2001)

Morton, H.V.: *The Heart of London* (Methuen, 1925)

Morton, H.V.: *The Spell of London* (Methuen, 1926)

Morton, H.V.: *London* (Methuen, 1926)

Morton, H.V.: *The Nights of London* (Methuen, 1926)

Pike, David L: *Subterranean Cities: The World Beneath Paris and London, 1800–1945* (Ithaca: Cornell University Press, 2005)

Roberts, Chris: *Cross River Traffic: A History of London's Bridges* (Granta, 2005)

Rose, Jacqueline: *On Not Being Able To Sleep: Psychoanalysis and the Modern World* (Chatto and Windus, 2003)

Sala, G.A.: *Gaslight and Daylight. With Some London Scenes They Shine Upon* (George Chapman and Hall, 1859)

Sala, G.A.: *Twice Round The Clock: Or, The Hours of the Day and Night in London* (Houlston and Wright, 1859)

Sante, Luc: *Low Life: Lures and Snares of Old New York* (New York: FSG, 1991)

Scanlan, John: *On Garbage* (Reaktion, 2005)

Schivelbusch, Wolfgang: *Disenchanted Night: The Industrialization of Light in the Nineteenth Century* (1983 and Berkeley: University of California Press, 1995)

Schlor, Joachim: *Nights In The Big City: Paris, Berlin, London – 1840–1930* (1991 and Reaktion, 1998)

Solnit, Rebecca: *Wanderlust: A History of Walking* (New York: Viking Penguin, 2000)

Sullivan, Robert: *Rats* (Granta, 2005)

Tardi, Jacques and Leo Malet: *The Bloody Streets of Paris* (1996 and New York: ibooks, 2002)

Tatsumi, Yoshihiro: *Abandon The Old In Tokyo* (1970 and Montreal: Drawn & Quarterly, 2006)

Walkowitz, Judith R.: *City of Dreadful Delight: Narratives of Sexual Danger in Late-Victorian London* (Chicago: University of Chicago Press, 1992)

Wallen, Martin: *Fox* (Reaktion, 2006)

Winlow, Simon and Steve Hall: *Violent Night: Urban Leisure and Contemporary Culture* (Oxford: Berg, 2006)

Wright, Patrick: *The River: The Thames In Our Time* (BBC, 1999)

Wyndham, Horace and Dorothea St. John. George: *Nights In London* (John Lane, 1926)

Yancey, Philip: *Prayer: Does It Make Any Difference?* (Michigan: Zondervan, 2006)

Night Haunts first appeared as a collaborative project on a website [www.nighthaunts.org.uk]. My creative partners, Ian Budden and Robin Rimbaud, made the process, which took well over two years, both fun and enlightening. Mark Diaper, the designer of this book, has made the words strange and new again. Mala Haarmann has generously supported *Night Haunts* from the outset.

Many individuals helped me during my nocturnal researches. Some of them, for reasons of shyness or through fear of retribution by their employers, have asked to remain anonymous. Those I can and do wish to thank: Pallavi Aiyar, Meena Alexander, Paul Bagott, Harinder Bains, Kiran Bains, Sophia Bains, Paul Bates, Eugenia Bell, Gary Bell, Ben, Richard Blint, Andrew Boyd, Bob Catterall, Alexandra Chang, Sarah Crompton, Paul Davies, Kodwo Eshun, Gareth Evans, Kieron Evans, Faction G, Eliane Glaser, John Gleason, David Godwin, Elizabeth Goff, Ruby Gomez, John Greenleaf, Madala Hilaire, Ed Hammond, Dean Harris, Peter Heims, Peter Hudson, Tony Hussein, Ian Jack, Alan Jenner, Janice Kerbel, Paul Laity, Penny McLean, Guy Meadows, Jamie-James Medina, Elizabeth and Vincent Meyer, Merope Mills, Hal Momma, Mary Morrell, Alan Murdie, Ali Murtaza, Paul Myerscough, Joel Oluwafemi, Norman and Yvonne Palmer, Luis Pamplona, Papa, Matt Parton, Tom Penn, Mark Pilkington, Tim Robey, Michelle Rubin, Anjalika Sagar, SF Said, Kulwant Sandhu, Pritam Sandhu, Karin Schulze, Bruce Lindsay Smith, Caspar Llewellyn Smith, Andrew Steer, Rebecca Strong, Rosie Swash, Zilda Tandy, Riccardo Tartaglia, Jack Tchen, Jenny Turner, Laura Wallace, Matt Weiland, Adrian Williams.

The staff at Artangel, particularly Sarah Davies, Maitreyi Maheshwari and Janette Scott, as well as Michael Morris, have been as helpful as they have been patient with me. James Lingwood has been unstinting in his encouragement and acute in his advice. Special thanks to Cathy Haynes who first showed me a copy of H. V. Morton's *The Nights of London* and convinced me it would be a good idea to give up sleeping for a couple of years. As injurious as those nights have been to my health, they have brought me closer to people and places, and allowed me to gain insights about human goodness and resourcefulness even in the most straitened of circumstances that, far from diminishing my love for London, have served to amplify it.

Artangel

Night Haunts was commissioned by Artangel with the generous support of Mala Haarmann. *Night Haunts* is part of the *Nights of London* programme, funded by Arts Council Lottery with the generous support of Vincent and Elizabeth Meyer and the John Lyons Charity.

Artangel's programmes are made possible through the generous support of Arts Council England, London and The Company of Angels.

Artangel is a registered charity no. 292976.

Supported by
The National Lottery®
through Arts Council England

www.artangel.org.uk